The Inheritor's Powder

Also by Sandra Hempel

The Medical Detective

The Inheritor's Powder

A Cautionary Tale of Poison, Betrayal and Greed

SANDRA HEMPEL

WEIDENFELD & NICOLSON
LONDON

First published in Great Britain in 2013
by Weidenfeld & Nicolson
An imprint of the Orion Publishing Group Ltd
Orion House, 5 Upper St Martin's Lane, London WC2H 9EA

An Hachette Livre UK Company

1 3 5 7 9 10 8 6 4 2

A CIP catalogue record for this book is available
from the British Library

ISBN (Hardback) 978 0 297 86742 5
ISBN (Export Trade Paperback) 978 0 297 87035 7

Typeset by Input Data Services Ltd,
Bridgwater, Somerset

Printed and bound by
CPI Group (UK) Ltd, Croydon, CR0 4YY

The Orion Publishing Group's policy is to use papers
that are natural, renewable and recyclable products and
made from wood grown in sustainable forests. The logging
and manufacturing processes are expected to conform to
the environmental regulations of the country of origin.

www.orionbooks.co.uk

CONTENTS

To Georgia and Sophie, the Nibbs ladies, with all my love

The Fell Spirit of the Borgias

On 17 December 1846, Saunders and Otley of Conduit Street, London, announced the publication of *Lucretia, or the Children of the Night*, a novel in three volumes. Its author, Sir Edward Bulwer-Lytton, MP, was one of the most popular and prolific writers of the time, whose colourful work included the celebrated opening line, 'It was a dark and stormy night'. Priced at 31s 6d, *Lucretia* was an immediate success, with many thousands of fans right across the social divide, 'in the drawing rooms of the aristocracy, in the parlours of the tradesman', said one newspaper.

But not everyone was impressed. *The Times*, though not too high minded to take the publisher's money for an advertisement, devoted two and a half columns to an emotional attack on Sir Edward's 'sickening and unpardonable revelations'. Only the *Newgate Calendar* contained so many murders in so few volumes, the reviewer wrote, while none was half so terrible and monstrous as those in *Lucretia*. Moreover, Sir Edward had 'dallied so long with crime and criminals, had thrown so sickly a halo around the forms of vice, had taken such pleasure in the tricking out of naturally repulsive thoughts, that we knew it to be impossible for the man to depart for ever [from writing fiction] without some crowning work of hideousness and strangely morbid fancy'. *Lucretia* was that work, 'a disgrace to the writer, a shame to us all'.

Lucretia is certainly, to paraphrase Sir Edward, a dark and stormy tale. Blood 'gushes and plashes', moonbeams move 'creepingly and fearfully' down an oaken staircase, a guillotined mother's 'trunkless face' oozes gore, and all the while the heroine Lucretia Clavering poisons her way from sensation to sensation. As something of a clue to her character, she shared a first name with Lucrezia Borgia, one of history's most famous poisoners, while her surname was borrowed from an Essex village which had recently been in the news as a hotbed of criminal poisoning.

What provoked *The Times*' response, however, was more than the isolated publication of a gruesome piece of popular fiction, for *Lucretia*'s appearance coincided with the outbreak of a poison panic among the British public that was to last well into the next decade. 'The fell spirit of the Borgias' was 'stalking through English society', wrote one commentator, and Sir Edward was accused not only of encouraging would-be poisoners, but also with providing them with a manual to guide them.

The paranoia of early Victorian Britain that saw poisoners lurking in kitchens and behind bed curtains throughout the land, their little bags of white powder at the ready, was fuelled by several notorious cases, including the strange story of an unremarkable old farmer in an obscure village some thirteen years before the drama of *Lucretia* …

The Big Square House in the Village

At six o'clock on the morning of Saturday, 2 November 1833 – All Souls' Day or the Day of the Dead – in a cottage at the end of the track that led through landowner George Bodle's orchards, young Mary Higgins came into the kitchen to start her chores. Mary was servant to George Bodle's son, forty-seven-year-old John, and his wife Catherine, and had been with the family for three years.

John Bodle, known to everyone as Middle John, was sometimes described as the manager of his father's estates, but the title is misleading, implying that he had an administrative, even a professional role. In fact Middle John was a labourer, sweating it out in the fields and the animal sheds alongside the hired hands just as his father had done before him. He lived in a house that his father owned and called in on his father every Saturday night to be handed his wages, just like the casual hands. Twenty-seven years earlier John had married Catherine Judd from the nearby parish of Lewisham. It was not a happy union – tales of infidelity and violence had given the village gossips plenty of fodder over the years – but in that November of 1833, for the time being at least, the pair seemed to have settled into a kind of shaky peace.

The household now consisted of John, Catherine, their two sons – twenty-three-year-old John (known as Young John) and twenty-six-year-old George – Mary Higgins and Catherine's nephew, a fifteen-year-old, rather simple lad called Henry Perks, who had lived with the family since he was a small child. Henry too worked for old George, as a cowherd and odd-job boy. John and Catherine also had a daughter, twenty-four-year-old Mary, who had married an older man, Thomas Andrews, from the nearby village of Charlton. The couple lived with their two children, George and Edward, in rooms over the coffee shop they ran in St John Street in the Clerkenwell district of London.

Mary Higgins had gone to bed at ten o'clock as usual the night before, leaving a small fire smouldering in the kitchen grate ready for rekindling the next morning. She was normally the first to be up and about in the cottage except when her master was in a hurry. Then he would call her before making his own breakfast and throwing more wood on the fire while she dressed. As a maid-of-all-work Mary was the lowest of the low in the servant hierarchy, doing duties that would be shared by the scullery maid, housemaid, parlour maid and laundry maid in a grander house.

By the middle of the century, a maid-of-all-work in London could expect to earn around £6 to £8 a year, with an allowance for tea, sugar and beer, but Mary's recompense would have been well below this and she may even have been glad to work simply in return for her bed and board in order to stay out of the poorhouse. Indeed, a Victorian manual on servants' duties stated that the maid-of-all-work was 'usually regarded as the hardest worked and worst paid of any branch of domestic servitude; it is, therefore, usually filled by inexperienced servants or females who are so circumstanced that they are only desirous of securing a home and of earning sufficient to keep themselves decently clad'. The unfortunate Mary was unable to achieve even that modest ambition: in January the parish had had to buy her a pair of stays for four shillings and in April they contributed again with a five-shilling bonnet. Later, after she had gone through a courtroom ordeal and found herself at the centre of a scandal, the overseers would give her a final ten shillings 'to relieve her things out of pledge and go to her father'.

Robert Kemp Philp, who published best-selling handbooks for aspiring housewives, set out the daily routine for all the Marys skivvying away in households across the country. 'The duties of a maid-of-all work being multifarious, it is necessary that she should arise early in the morning; and six or half-past six o'clock is the latest period at which she should remain in bed. She should first light the kitchen fire, and set the kettle over to boil; then she should sweep, dust and prepare the room in which breakfast is to be taken. Having served the breakfast, she should, while the family are engaged upon that meal, proceed to the various bedchambers, strip the beds, open the windows, &c. This done, she will obtain her own breakfast, and after washing and putting away the things, she will again go upstairs and finish what remains to be done there.

'As the family will in all probability dine early, she must now set about the preliminaries for the dinner, making up the fire, preparing the vegetables, &c. After the dinner is cleared away, and the things washed and put by in their places, she must clean the kitchen; and this done, she is at liberty to attend to her own personal appearance, to wash and dress herself, &c. By this time the preparation for tea will have to be thought of, and this being duly served and cleared away, she must employ herself in needlework in connection with the household, or should there happen to be none requiring to be done, she may embrace this opportunity to attend to her own personal necessities. Supper has then to be attended to; and this finished, the maid-of-all-work should take the chamber candlesticks, hot water, &c., into the sitting-room, and retire to rest as soon as her mistress or the regulation of the establishment will permit her.'

Philp conveys well the mindless drudgery of the life but fails to describe the considerable physical demands. Lugging buckets of water and baskets of logs; scouring the range, the oven and the floors; beating mats and doing the laundry were back-breaking jobs. And the pleasant domestic scene that Philp paints, with the servant 'employing herself in needlework', and the reference to 'the regulation of the establishment' were not quite how life was lived under the roof of John and Catherine Bodle.

On the morning of 2 November when Mary came downstairs a large log was already burning in the grate and a figure sat motionless by the hearth. The shutters were still closed against the dark and the candles unlit. It was not, however, Mary's master Middle John who was seated there, but his son Young John. Mary then remembered having heard the sound of someone chopping wood while she was still in bed. The young man told her that it had been him – a chip had flown up into his eye.

Slim with dark brown hair, a sallow skin and hazel eyes, Young John was an attractive man despite the scattering of small pits on his face, common at a time when diseases such as smallpox were prevalent. Charming as well as good looking, he presented an altogether more refined figure than his father and grandfather and the working country people around him, and his friendly nature meant that most of Plumstead was happy to forgive his aversion to work and his vanity over his appearance.

That morning John seemed taken aback at seeing Mary, the girl thought. 'He said, "What did you get up so soon for?". I asked, "What is it o'clock then?" and he said it was six.' She might have asked why he was surprised, for she was up at her usual time and until a few days before Young John had seldom been out of bed before midday. Two years earlier he had tried his hand at running a coffee shop like his sister in Shoreditch in London's East End, but the business had failed within months and he had come back to Plumstead. Like the rest of his family, he had originally been set to work by his grandfather, labouring on the farm alongside his father and brother, but this was hardly the life the young man had in mind for himself. In any case, that arrangement had ended abruptly one day in 1830 when something had happened to cause George to turn the boy out of the fields and ban him from the farm. Middle John had no idea of the reason – George refused to discuss it – but he bitterly resented the situation and itched to make his son take off his fashionable jacket and roughen the skin on his soft hands. In the meantime Young John, untroubled by his father's opinion, lived the life of a gentleman, with free food and board and handouts from his mother from the housekeeping, although one or two people suspected he might have another, undeclared, source of income.

Mary said nothing more but trotted off to start on her other jobs. When she came back half an hour later Young John was still in the chair before the fireplace, the shutters still shut. He asked her whether it was day yet; she told him no.

Meanwhile, up at the farmhouse along the track from Middle John's cottage, a less orderly state of affairs prevailed. There another maid-of-all work, Sophia Taylor, who had been working for old George and his wife Ann for nearly three years, was running late with her duties. The nineteen-year-old, a somewhat more feisty character than Mary Higgins, had had her 'holyday' the day before and spent part of it visiting the Baxter family 5 miles away at Ruxley Farm in the hamlet of North Cray.

Samuel Baxter had married George Bodle's daughter, Mary-Ann, and was held in high regard by the old patriarch. When George and his wife Ann died, Samuel and his eldest son William were due to inherit substantial amounts of land and income, and there were legacies to the other Baxter children as well as investments that the old man had left to Mary-Ann personally. William Baxter was a few months younger than his cousin Young John but a very different man, already farming in his own right. The family had moved to Ruxley Farm in 1824 from Plumstead, where Samuel had been renting a large house and several parcels of land, including one of George's fields.

Gossip had it that Sophia Taylor was trying to steal the eldest Baxter daughter's sweetheart, John Wood, and indeed Wood was rumoured to have exchanged Louisa

Baxter's company for the pleasure of walking Sophia the 5 miles home that Friday night. Certainly by the following September Wood's romance with twenty-one-year-old Louisa was over, for the young woman married a Thomas Wilkes of Woolwich. Whatever the truth of the story, Sophia had obviously spent a more exciting day than Mary Higgins, and it was past seven before she was up, dressed and in the kitchen, clearing out the grate and making up the fire.

Once the fire was lit it was fifteen-year-old Henry Perks' job to fill the great iron kettle from the pump in the yard and heave it on to the metal crane in the fireplace. Henry, who lived with Young John down at the cottage, usually reached the farmhouse at about six. As well as filling the kettle he would let out the hens and guinea fowls, and beat the mats in the yard. But like Sophia, Henry was late starting his chores that morning so he abandoned the kettle on the back step and set about the more important job of fetching the cows.

Dawn breaks at just before seven in south-east England in early November. Back at the cottage Mary Higgins heard Young John pick up the milk can from the storeroom near the back door and set off at a run along the path to his grandfather's house as the sky began to lighten. It was a mild day for the time of year, and the farmhands starting their day's labours called out to him and he shouted a cheerful 'Good morning' as he jogged by, swinging his can.

Young John had been collecting the milk from his grandfather's house for about a fortnight but he had never before left his bed so early. The visits had started as a joke between him and Sophia, the girl encouraging him, he said, with flirty invitations and the delightful prospect of joining her in the cellar while she churned butter. He thought he had been invited to breakfast the previous Monday and had turned up at six o'clock with some tea and, appropriately as things were to turn out, a red herring. Sophia, however, had sent him away, for this was the start of what was known as washday, an event that took place on average every five weeks and, despite its name, occupied four people for the best part of three days. Mrs Lear had arrived earlier than usual in readiness and the copper had been filled from the tap in the yard and put on the fire to heat.

John had come back the next day when the washing was well under way, bringing with him some more fish, and this time John, Sophia and Sophia's young helper, Betsy Smith, sat down to eat together, while Jeremiah Febring hung the latest batch of laundry out to dry on the line in the yard.

There was no such fun at the farm on this Saturday morning, however. When Young John put his head round the kitchen door he found Sophia in an irritable mood, cleaning the stove and grumbling about being behind with her work. He stayed just long enough to do the job that Henry had abandoned, filling the kettle in the yard and lifting it on to the crane over the fire. Then, at about half past eight, after Young John had gone, the household pecking order for breakfast began.

The coffee was kept locked away in the parlour, in a wide-mouthed glass jar of the sort normally used for bottling fruit. The old farmer came down with his keys to open the cupboard, measured out the daily coffee ration into a teacup and locked the bottle away again, as he did every morning. The servants were not allowed keys but Ann Bodle had a set and, like her husband, carried them about with her. Today supplies were running low: there was just enough coffee for that day left in the jar. The sugar was in the cupboard next to the coffee but with the price of sugar at only five pence and three farthings a pound compared with coffee at £1 7s, the Bodles didn't bother to guard it quite so well: the sugar was locked away only at night.

That morning supplies of the family's usual dark brown muscovado (known then as moist sugar) were also about to run out, so Sophie Taylor gestured to Betsy to go to her bedroom and fetch some of the lump sugar from the store that her mother had given her when she first went into service with the Bodles. Betsy was Ann Bodle's granddaughter through her first marriage and therefore old George's step-granddaughter; a good-looking, intelligent girl who was deaf and dumb. She lived at the farm with George and Ann but, unlike Young John, she worked for her keep, helping Sophia with the chores.

Betsy duly came back with some lump sugar in a bowl and while the girls prepared the farmer's simple breakfast, George went out to feed the fowls. He seemed quite well; he was taciturn but then he always was a man of few words. Sophia made toast and a large pot of coffee with the water from the kettle. The pot was quite clean and was never out of Sophia's sight. No one else touched it. The girl poured her master a cup of coffee, added milk from a jug that Betsy had brought earlier from the cowshed, and some of the last of the thick, treacly sugar. As George breakfasted alone in the kitchen, his wife Ann, who was seventy-four and very frail, had her breakfast in bed, carried up on a tray by Elizabeth Evans, a daughter by her previous marriage. Elizabeth often stayed at the farm to look after her mother.

Mrs Bodle had a small cup of coffee that morning sweetened with some of Sophia's lump sugar, and her husband had his usual half-pint bowl. About half an hour after the master and mistress had finished their breakfast, Elizabeth, Sophia and Betsy sat down to theirs. The three women topped up the coffee pot with more water from the kettle and reboiled it. Sophia had two cups. Betsy took hers without sugar, while Sophia and Elizabeth finished up the last of the Bodles' muscovado. The coffee grouts were then left in the pot in the wash-house sink as usual, ready for a third recipient to collect.

And so began a chain of events that was to enthral the country for weeks, filling sheet after sheet in newspapers across the country, from the *York Herald* to the *Leicestershire Chronicle* and the *Royal Cornwall Gazette*. The drama made such an impression on one local man that fifty years later he recalled: 'Fireside stories which we hear as children cling to us through life with the weird fascination of fairy fiction and, great and manifold as our own experiences may be, hardly any events passing within our knowledge can produce remembrances so deep and lasting as the early impressions on our tender minds with a stamp too deep to be effaced. To this day I cannot look without a shudder at the big square house in Plumstead village.'

2

A Great Chain and
Dependency of Things

'Those things which are experienced to be in their whole nature or in their most remarkable properties so contrary to the animal life as in a small quantity to prove destructive to it are called poisons,' wrote the early eighteenth-century physician Richard Mead. Mead, who included Queen Anne, George II and Sir Isaac Newton among his patients, was particularly interested in poisonous snakes and he dissected vipers in order to understand the mechanism of their fangs. In the best tradition of researchers, he also drank snake venom himself in order to prove that it was only harmful when delivered through a puncture wound. Fortunately, he was right about that.

What we know as the modern science of toxicology has evolved into so much more than the simplistic definition of the study of poison. The US Society of Toxicology's less snappy preference is: 'The study of the adverse effects of chemical, physical or biological agents on living organisms and the ecosystem, including the prevention and amelioration of such adverse effects.' These effects can range from virtually instant death, as in cyanide poisoning, to subtle changes in the body that take years to manifest themselves. They also include diseases caused by toxin-producing bacteria, such as diphtheria and cholera.

Mead's fascination with vipers was partly inspired by their ancient reputation as instruments of divine vengeance, punishing mankind for infringements of God's law. In fact, the mysterious, devastating nature of any kind of poisoning was often seen as requiring the services of the priest or the magician rather than the doctor. At the same time Mead pondered how a benevolent creator could allow such destructive substances to exist in his world.

Mead was not the first to think about the nature of poison. Man's early understanding that some of the plants and animals around him were harmful to swallow or touch led, predictably, to the harnessing of that knowledge for the purposes of war and murder. Native North Americans, Africans and South American tribes dipped their arrows in whichever toxin was most readily to hand: snake venom, sweat from poisonous frogs or concoctions derived from poisonous plants. 'For the arrows of the Almighty are within me, the poison whereof drinketh up my spirit,' wails Job in the Hebrew Bible, while Homer sends Odysseus on a voyage to obtain 'a deadly drug that he might have the wherewithal to smear his bronze-shod arrows', and Ovid recounts that after Hercules had killed the many-headed Lernaean serpent he dipped his arrows in its venomous blood. An ancient Egyptian text threatens anyone who betrays its sacred secrets with 'the penalty of the peach', implying that the priests knew how to extract cyanide from peach kernels. In 399 BC the Greek philosopher Socrates, sentenced to death for corrupting the minds of young people and for impiety, was given hemlock to drink. The numbness slowly crept up his body until it reached his heart.

The Ebers Papyrus, an Egyptian medical treatise written around 1,500 BC but probably based on texts written nearly two thousand years earlier, refers to many common poisons, including hemlock, aconite, opium, lead and antimony. The ancient Hindu Ayurveda texts recount how poison came into being: 'Soon after the creation of the world, Bramha was displeased with Kaitaba, one of the demons, and in his anger poison was generated.' After listing pages of vegetable, mineral and animal poisons with their various symptoms and antidotes, the treatise gives advice on how to spot a poisoner: 'He does not answer questions or they are evasive answers; he speaks nonsense, rubs the great toe along the ground and shivers; his face is discoloured; he rubs the roots of his hair with his fingers and tries by every means to leave the house.'

In the twenty-seventh century BC Shen Nung, known as the father of Chinese medicine, is said to have died as a result of writing his treatise 'On Herbal Medical Experiment Poisons', after tasting 365 different herbs for his research. Around 2,300 years later, Shen Nung's Western counterpart Hippocrates discussed the concept of overdose, and through his understanding of the need to limit the absorption of poison in the gut, he introduced what we would now call the concept of bioavailability: the extent to and rate at which a drug is available at the site of its action in the body. The route by which a poison gets into the body – by mouth, lungs, skin or injection – is crucial in determining its strength and speed of action.

Later, Greek and Roman physicians added to the body of knowledge. Nicander of Colophon, the second-century Greek poet and physician, wrote long verses on the subject of venomous animals and antidotes. He is said to have gained his knowledge of white lead, red lead oxide, aconite, cantharides, hemlock, henbane and opium through experimenting on criminals. In AD 65 Nero's doctor Discorides began classifying poisons according to whether they were animal, vegetable or mineral.

History's most famous poisoners, the Medicis and the Borgias in fifteenth-century Italy, almost certainly used arsenic but they were well ahead of the game. Arsenic poisoning was impossible to diagnose until the late 1700s, and even then the process was distinctly hit or miss: medical science was unable to distinguish the symptoms from those of common illnesses such as food poisoning and dysentery. Even if arsenic did come under suspicion, there was nothing resembling a reliable test to confirm its presence.

Again in Italy, two centuries after the Borgias, there were rumours of a mysterious slow poison that was thought to have magic powers, allowing a murderer to bring about the victim's death whenever he chose, after months or even years of gradual decline. Eventually a woman called Giulia Tofana was accused of selling the potion to women who wanted to dispose of their husbands. Amid mounting public hysteria, Tofana eventually confessed to bringing about the deaths of at least six hundred men, but the admission was obtained under torture and by then truth and rumour had become impossible to untangle. Certainly there was a spate of poisonings in Rome and Naples around that time, but the active ingredient in the supposedly weird and wonderful

Aqua Tofana was probably nothing more supernatural than white arsenic.

A wandering sixteenth-century Swiss physician, alchemist and astrologer, Auroleus Phillipus Theostratus Bombastus von Hohenheim, better known as Paracelsus, explored the subject through his interest in the medicinal properties of metals and minerals. While he believed that 'nature hints at cures', Paracelsus began looking for alternatives to what he saw as the often ineffective herbal remedies that then dominated medicine.

Feared and admired as an evil magician, a charlatan and a genius, and dubbed both the Martin Luther of medicine and the reincarnation of Dr Faust, Paracelsus bridged sorcery and science. His thinking about the difference between a therapeutic dose and a poisonous dose for each of the substances he investigated led him to identify what we now know as the 'dose-response relationship', now a key aspect of toxicology. His maxim 'All things are poison, and nothing is without poison; only the dose permits something not to be poisonous' still holds good. Paracelsus also tried to identify which chemical was responsible for the toxicity of a particular plant or animal, while his theory that diseases tended to concentrate their effects on a particular part of the body resulted in the idea of a poison's 'target organ'. His work marked the first step in moving the study of poisons into something approaching a science.

But Paracelsus apart, there would be no major breakthroughs in the understanding of poisons until the nineteenth century.

In 1811 Benjamin Brodie, who, like Richard Mead, was

to become a doctor to the royal household – George IV, William IV and Queen Victoria in his case – published the first of two studies of everyday poisons such as alcohol, tobacco and arsenic, as well as some more exotic substances such as upas antiar, used in Java for poisoning arrows, and curare, or woorara as Brodie called it, from South America. Dozens of dogs, cats, rabbits and guinea pigs were tortured for the purpose as Brodie poured poison down their throats, inserted it into their rectums and rubbed it into wounds he had made on their bodies. He took detailed notes of their dying symptoms and then dissected their bodies to see the internal effects. One of his accurate conclusions was that arsenic had to enter the bloodstream in order to do its damage. Brodie's work was quickly overtaken, however. Three years later, the first in what would be a series of comprehensive, scientific toxicology studies became a best-seller.

Richard Mead finally came to the conclusion that the question of how a loving God could allow poison to exist was to do with the nature of poison, which was not as simple as it might seem. Some poisonous plants, correctly prepared and prescribed, were medicinal and even provided food for other animals: 'Goats and quails are fattened by hellebore; starlings by hemlock and hogs innocently eat henbane.' And while the benefits of poisonous minerals were not obvious, arsenic, to take one example, was an active substance 'made use of by nature in preparing several metals in the earth, which are of great service to mankind'.

'In short,' Mead said, 'there is in the fabric of the world a great chain and dependency of things one upon another, and though our knowledge does not reach to every particular link of it, yet the farther we advance in the study of nature, the more we shall find it.' So it seemed that even the most toxic of animals, minerals and plants had their rightful place in the great scheme.

3

Death by Toad or Insect

'A scene of desolation is fast creeping around us; and the winter, whether severe or not, will probably be long and the diminution of comfort for man and beast thereby in some degree extended,' a melancholic *Maidstone Gazette* had warned as that October of 1833 gave way to November. The amateur forecasters trying to predict the coming season for the part of Kent where Bodle's farm lay were finding their task more confusing than usual. The outgoing month had been mild and mostly dry, just broken once or twice by heavy rain and thunderstorms, but oddly the swallows, swifts and sand martins had left sooner than usual while the incoming winter birds were early. The movement of the birds and an early leaf-fall seemed to signal a cold winter, yet the hedgerows were missing the mass of berries and wild fruits said to be a sign of harshness to come.

But while the weather figured large in country people's lives, it wasn't the only matter of current interest. For one thing, the habitual offender Henry Simmons was on the run. Given into custody by the Reverend Moneypenny on suspicion of stealing a wooden door, value 6s, the property of one Elizabeth Cheeseman, Simmons had escaped the clutches of Constable Walter Barton while being escorted to Maidstone county jail, and the authorities had announced a two-guinea reward for his capture. The public was told to look out for a round-shouldered individual with thin sandy hair, a round, 'frackley-smock' face, dirty complexion and an overhanging upper jaw with a front tooth missing. The alleged miscreant was last seen wearing a skin cap, two pullover smocks, a red waistcoat and barragan trousers, all very dirty; bad shoes – and handcuffs.

And if this were not excitement enough, Madame Panormo and the self-styled Infant Prodigy Miss Wildman Gould were to give what was billed as a vocal entertainment at the Star Inn Assembly Rooms, Maidstone. The programme would include a speaking ballet called 'Mama Declares that I'm too Young' and a song by Madame to be performed 'in character' entitled 'I'm a Brisk and Sprightly Lad'. Meanwhile, the local agricultural associations were preparing for the autumn ploughing, hop drying and lambing competitions, the Kent Auxiliary Bible Society was congratulating itself on a satisfactory annual meeting, Messrs Clark and Evans announced the sale of gunpowder tea 'in its genuine state' direct from the East India Company's London warehouse, and the price of wheat dropped to £2 7s 3d a quarter at the county market.

On the evening of Saturday, 2 November, just as Madame Panormo and the Infant Prodigy were preparing to take to the stage, a two-seater phaeton rattled along a country road in the faint light of a waning moon. Thirty-four-year-old John Butler, a partner with his father John senior and his younger brother Ebenezer in a respected Woolwich family medical practice, was on his way to Bodle's farm in response to an urgent message.

George Bodle's flat-fronted ten-room brick house stood on the main road through Plumstead village, an imposing landmark among the ramshackle cottages and pigpens which signalled to passers-by his prosperity and standing among his neighbours. Like its owner, the building was solid and comfortable rather than elegant. A three-chimney stack rose at each end of the pitched slate roof and a path ran from the front gate up to the central front door, on either side of which were two tall sash windows. These were topped by a duplicate set of windows on the first floor, with a fifth window centred over the front door. Usually, though, family and visitors alike entered through the yard at the back of the house, past the hen cages and the wash-house.

A cart track led down the side of the building, through the market gardens and the apple and pear orchards at the back, then bent in a large right angle, cutting through the lines of fruit trees before skirting a cottage and ending at a gate into the public way called Skittles Lane (although the locals called it Kiddels). Bodle's orchards continued on the south side of the track, culminating in a gravel pit and a common with an old mill where the local housewives took their small stores of wheat. The main road to the west between the farmhouse and the nearest dwelling, the St Nicholas vicarage, was fronted by Bodle's bullock sheds. Opposite the house, on the other side of the high street, the farmer's crops rolled down towards the river, giving an uninterrupted view across to the Essex countryside on the far bank of the Thames.

In his last years the old man himself, physically frail now

but as firm as ever in mind and purpose, could be seen pottering around the front of the house or standing at the gate, watching over his domain. After a lifetime of thrift and hard work, George Bodle had transformed himself from a tenant farmer into a major landowner and a man of substance.

The charwoman, a forty-nine-year-old widow called Judith Lear, had found a sorry scene when she arrived at the farm that morning, however. Every day after breakfast Mrs Lear came to the back door to collect some milk and the Bodles' coffee dregs for her daughter's family. Mary was married to a local farm labourer, Daniel Bing, and the pair lived in a tiny cottage on subsistence wages with their seven children, aged from twelve years to two months. Such was their poverty that Mary would boil up what was left of the Bodles' coffee after it had already been boiled twice – once for George and Ann Bodle and then for Elizabeth and Sophia – to make a hot drink for her children.

That morning Mrs Lear found the coffee pot standing in the sink in the Bodles' wash-house as usual, but she also found Sophia and Betsy, pale and ill, with stomach pains, vomiting and with a burning sensation in their throats. Sophia was particularly bad and had vomited violently several times. After sympathising with the young women, Judith Lear tipped the coffee grouts from the bottom of the pot into her jug and set off along the main Plumstead road towards her daughter's house, where she left the coffee on the table. No sooner had she got back to her own tiny cottage, however, than a tottering Betsy arrived to summon her back to the farm, indicating in sign language that Mrs Bodle needed her urgently. Mrs Lear found the old lady in her bedroom, clearly ill but still struggling to get dressed.

Despite the charwoman's help, in the end Ann proved too weak to walk down the stairs.

After settling her mistress back in bed, Mrs Lear went down to the kitchen, where she found George Bodle sitting at the table in as bad a state as his wife. 'He couldn't think what had occasioned it,' the charwoman said. 'He said he had eaten nothing for supper but a roast potato and had only coffee and toast for breakfast.' And there was something else: he said his eyes were very dim. He wondered whether something might have been wrong with the water used to make the coffee. Mrs Lear thought perhaps an insect or a toad had got into the kettle. George didn't know but, strangely, he said he was sure there was nothing amiss with the coffee because no one had had access to it apart from himself, and he asked who had been to the house that day. He then told the charwoman to make sure that the kettle was thoroughly scoured and scrubbed before it was used again and Mrs Lear set young Henry Perks about the task; he used a chisel to chip away at the build-up of limescale. At noon Elizabeth Evans, who had set off after breakfast for her home in the nearby hamlet of Bostall, staggered back into the house and collapsed. Two of the farmhands had to carry her up to bed.

At around six o'clock that evening the old man's son Middle John arrived at the farmhouse to collect his week's wages and found his father still in the kitchen, where he had been sitting all day, still extremely sick. Middle John took his money and went upstairs to enquire about his stepmother Ann before going back to his cottage to tell his wife the news. Catherine put on her bonnet and walked up to the farmhouse to see how the family was faring. For some reason the old man refused to speak to her, but rather pointedly took off his gaiters and went upstairs to his bedroom as soon as she entered the kitchen. After Catherine had gone, he came back down to await the doctor.

Careful with his fortune though George Bodle may have been, this delay in calling a doctor wasn't unreasonable in 1833. 'English cholera', so called to distinguish it from the deadly Asian cholera which had killed 32,000 people in Britain two years earlier, was a general term for the unspecified attacks of sickness and diarrhoea, usually caused by bad food, that commonly struck people down in the nineteenth century.

'The English', though unpleasant and debilitating, usually cleared up inside a few hours without the need for medical help, although it did occasionally prove fatal if the attack was especially vicious or the victim frail. Even when a medical man was sent for, it was far from clear whether his ministrations did anything to help. Fortunately most stomach upsets were short lived and self-limiting. What the doctor found when he reached the Bodle household, however, was on a different scale.

When John Butler began examining his patients and heard their tale he was struck by the similarity of the cases; not just the symptoms but the way everyone had fallen ill within minutes of having breakfast and the fact that they had all consumed the same toast and coffee. The old man was particularly affected: as well as pain, weak eyesight, vomiting and diarrhoea, his mind seemed disturbed. 'I found him exhausted, languid and feeble,' the doctor said. 'He appeared to be labouring at the same time under great anxiety and his intellects seemed to be much impaired. He did not seem stupid but exhausted.' As a result, Butler's medical prognosis was not encouraging: 'I was of the opinion that an old man would not rally or recover from it.' And the more he saw and heard, the more firmly Butler dismissed any suggestion of insects, amphibians or cholera of any nationality. The Bodles were, he was sure, suffering from the effects of an irritant poison, and at the top of his list of suspects was arsenic trioxide, more popularly known as white arsenic.

A lump of the element called arsenic may pass perfectly safely through the human body provided it remains unchanged in that elemental state. The chemical compound white arsenic is an altogether different proposition. In this form, the harmless arsenic becomes a rapid and horrible death sentence for all animals with a central nervous system and most plant life unfortunate enough to absorb it.

Arsenic is present everywhere: in the crust of the earth, in space, in the sea, in spring water, in mountains and also, in tiny traces, inside the human body. In the natural world it is usually found in combination with other elements such as sulphur and iron, as in the silvery-white arseno-pyrite, the orange-red realgar and the golden-yellow orpiment, all of which are highly toxic. The word itself comes from *arsenikon*, the ancient Greeks' name for orpiment, which they in turn took from *zarnikh*, the Persian word for yellow. *Arsenikon* is also related to the Greek word *arsenikos*, which means male or potent. The Egyptian boy king Tutankhamun was buried along with a linen bag coloured with orpiment, while the red hues of the pottery excavated at Corinth were found to derive from realgar. Such is the beauty of their deep glowing colours that even after the dangers were discovered, realgar and orpiment were still used for centuries as pigments in paintings, textiles and cosmetics, and to decorate buildings. The ninth-century illuminated Celtic gospel, the Book of Kells, and the Taj Mahal are both decorated with orpiment.

Because of its intense yellow colour, orpiment was thought to contain gold, and the *Mappae Clavicula*, a medieval compendium of ancient recipes, gives instructions on using the compound to extract gold, to gild iron and to make silver from copper. The Roman emperors Caligula and Diocletian are both said to have set their experts to work on this particular route towards the age-old quest, with Diocletian flying into a rage and destroying all the books on the subject after his Egyptian alchemists failed to fulfil their promises. The Roman statesman Pliny refers to 'a recipe to make gold from orphiment which occurs near the surface of the earth in Syria and is dug up by painters', while the Greek geographer and historian Strabo described a mine of yellow and red sulphides in Pompeiopolis that was so poisonous that only slaves were employed to work there. In nineteenth-century Britain, however, everyone's favourite deadly pigment was neither red nor yellow but a rich, vibrant green.

In 1775, a Swedish chemist had developed a colour that was then named after him – Scheele's green or copper arsenite, a compound of copper, arsenic, hydrogen and oxygen. Then in the 1800s Scheele's was replaced by what was seen as a better product, longer lasting and with a wider range of shades. This was an equally toxic variant on Scheele's called Emerald or Paris Green, not because it was seen as fashionable and sophisticated – although it was – but for the rather less romantic reason that it was used to kill rats in the sewers of the French capital. Soon green pigments containing large amounts of arsenic were everywhere – in paints, wallpaper, fabrics, soap, toys, sweets, cakes and candles – making it hard to avoid touching, inhaling or swallowing the stuff in one form or another. In his work on occupational diseases, the doctor John Arlidge refers to artificial flowers containing on average 10 grains of arsenic while 20 yards of material in a green ball dress were found to harbour 100 grains. The famous claim that Napoleon, living in exile on St Helena, was poisoned by his wallpaper is based on the theory that he was exposed to arsine gas given off by the green pigment.

But it is arsenic trioxide, which nineteenth-century chemists called arsenious acid and everyone else called white arsenic, that most people mean when they refer simply to arsenic. This form is a by-product created when ores are smelted to extract valuable metals such as copper, lead and gold. As the ore is heated, the elemental arsenic, which is naturally present in large quantities, is given off as a gas. It then combines with oxygen in the air to form a substance that might have been purpose-designed for the easy disposal of inconvenient relatives.

The eighth-century alchemist Jabir ibn-Hayyan is credited with introducing white arsenic to the world when he heated either realgar or orpiment – accounts vary – and then collected the substance that formed when the vapour hit the air. Jabir's book of poisons lists some exotic recipes, including the directions: 'Take a gecko and a yellow tarantula, then pulverise them both finely; they are mixed with milk and left there to ferment.' But it was his rather more prosaic-sounding contribution to toxicology that turned out to be both more enduring and more dangerous, for along with the discovery of white arsenic came the knowledge that was to keep British doctors, chemists, lawyers and juries, not to mention hangmen and undertakers, busy for over a hundred years. For unlike the violently coloured orpiment and realgar, arsenic trioxide is the most innocuous looking of powders, tasteless, odourless, resembling at a quick glance flour or sugar, and easily dispersed in hot food and drink. It is also fatal in tiny doses, and in Britain in 1833 it was cheap and ridiculously easy to get hold of, a situation that resulted in no end of mischief ...

4

That Good and Pious Man

Like most of the extended Bodle family, Samuel Baxter was in and out of the farmhouse on a regular basis. Since his marriage to George's daughter over twenty years earlier, he had become more of a son than a son-in-law to the old man, confided in and trusted in a way that Middle John was not. Only a few days before George fell ill, Samuel had helped the old farmer to draw up a new will. Baxter was summoned to the house on the afternoon of Saturday, 26 October, where he said he had found the old farmer sitting alone, clutching a sealed packet. George handed the package to Samuel and asked him to take it immediately to his solicitor, Charles Parker, in Greenwich. Samuel claimed that at the time he had no idea what the packet contained; George did not tell him and he did not ask.

Five days later Samuel was back in Parker's office but he was accompanied by George and this time he certainly knew what was in the packet. There in the presence of witnesses, Parker's clerks Alex Grove and Thomas Ashdown, George Bodle signed his new will. Rumour had it that the entire Bodle family knew both the contents of the old will, which was then destroyed, and of the new, but if so they were keeping the information to themselves. Middle John was later to say he believed that under the old will he had been left everything, while Samuel would deny ever knowing what the previous document had contained. Either way, what every member of the family did know was that old George Bodle was a wealthy man.

Born in the village of Plumstead in 1754 to Willum and Mary Bodel, as the parish register had it, George was one of three boys and three girls, but deaths in infancy and childhood were commonplace then and few couples in the village saw all their children survive into adulthood. George's sister Eleanor died in 1763 aged just five weeks, followed three years later by sixteen-year-old William junior.

George grew up to be a quiet, undemonstrative man, plain mannered and plain living. Despite his steadily growing wealth, he never aspired to gentility but continued to dress in the rough clothes of the practical farmer and to follow the old country ways. He was a labouring man who worked his fields alongside his hired hands until his strength failed him; a man who valued probity, believed in order and feared God.

His first wife, called Mary like his mother, was ten years younger than he was and bore him at least three children – John, Mary-Ann and George, their firstborn, who died in 1790. The couple almost certainly had at least one other child, a boy called William, who probably died in childhood too. Mary died in 1799 at the age of thirty-five, the cause of her demise not recorded.

Nine years later, on 29 February 1808 at St Mary's church

next to Lambeth Palace, George married again. This time his bride was not a local woman although she did have family ties with Plumstead. She was a Lambeth widow called Ann Wassell, a relative of one of George's friends, the wealthy George Wassell, from nearby Woolwich.

In the meantime George had prospered. In 1786, aged thirty-two, he had taken out a policy with the Sun insurance company in the City of London that had valued his house at £300. He also listed a barn worth £200, equipment and stock valued at another £150, two stables, a cow house, a cattle shed, a granary and a cart lodge. Household goods accounted for another £100 and 'wearing apparel' for £50, but his most valuable asset was the great brick-built stack yard with its stores of hay and wheat. And as well as the insurable assets, there were his investments and the land, for George was also a tenant-occupier of large swathes of the fertile drained marshland that lay between the village and the banks of the River Thames.

On the stretch from the convict wharf to the crab-tree sluice, known as the crab-tree level, he was renting eight acres and on the next strip, the old corn-marsh level, he was the tenant of another 52 acres. Steadily over the years he managed to acquire the freehold to 24 acres of that land and in turn began leasing it out to other farmers. He was also dealing in thoroughbred horses. In 1823 at a magistrates' hearing at the Mitre Tavern, Greenwich, he gave evidence against a fellow dealer, James Haines of nearby Lewisham. Haines had bought a blood mare from him for 10 guineas with what turned out to be counterfeit coins.

By November 1833, George Bodle, 'that truly good and pious man' (as his tombstone later had it), possessed some

40 acres of freehold land along with the cottages and barns that stood on them; leasehold property covering a considerable slice of the village; a farmhouse with adjoining orchards, market gardens and outbuildings; livestock and grain stores; and £3,000 of stock at the Bank of England. In all, he was worth about £20,000 (slightly over £2 million in 2012 terms).

If Middle John had been right about the old will leaving everything to him, then the new one must have come as something of a disappointment. Consisting of more than three pages of thick, densely packed black script, the words running together almost continuously in the style known as cursive, it is a detailed piece of work. The old farmer had a fair amount to leave and gave intricate instructions about what was to happen to the money and property he had amassed so cannily over his lifetime.

After directing, according to legal custom, that any debts outstanding on his death be paid out of the estate, along with funeral expenses and the costs of proving the will, the old man gives to his widow 'all my wines, liquors and household stores absolutely, also to her my said wife for her life or while she remains and continues my widow and unmarried but no longer, the yearly rents and profits of all my freehold estates and premises with their respective appurtenances … and also the interest of all my money in the public stocks or funds of this kingdom. And likewise the free use, wearing and enjoyment of all my household goods and furniture and plate, linen, china, books, pictures, prints, glass and household effects of every description and likewise the yearly income of all the residue and remainder of my goods, chattels, credits and estate and effects.'

Characteristically, while bestowing all of his personal possessions and the income from his considerable accumulation of land and investments on his widow, the last will and testament of the unsentimental George Bodle is devoid of any of the expressions of affection frequently seen in such documents. Ann is neither his 'beloved wife' nor his 'dearest wife' but simply 'my wife Ann Bodle'. Nor is anyone to receive a mourning ring as a token of the old man's esteem, a popular gesture at that time.

It was after Ann's death, or in the highly unlikely event of the sickly seventy-four-year-old woman remarrying, that matters became interesting. Middle John was to inherit the farmhouse as well as the cottage in which he and his family were currently living, together with the 13 acres of market garden and fruit orchards adjoining the farmhouse and 3 acres of marshland at the old corn-marsh level. This entire inheritance was only for the duration of Middle John's lifetime, however: he was entitled to the rents and profits from the land and buildings, but he could not sell them. On his death, everything was to be divided equally between his children – Young John, George and Mary Andrews. And unlike Middle John, the old farmer's three grandchildren were free to do as they wished with their share.

But old George Bodle was far from finished, and what followed came as news to Middle John – or so he claimed. The will continues: 'After the decease or marrying again of my said wife, whichever shall first happen, then I give and devise unto my son in law Samuel Baxter all my freehold messuages [a dwelling house with any adjacent buildings and land] or tenements [permanent property such as land or houses] with stable, cow-house, yard, garden and premises with their appurtenances situated at Plumstead aforesaid, now in the occupation of Mr Hore.' And in addition Samuel was to get 11 acres of freehold marshland at the corn-marsh level, adjacent to Middle John's 3 acres, and the crab-tree level, along with the adjoining stables, cow-byre, yard and garden that were currently rented out to a John Taylor. And unlike Middle John, Samuel would not be holding his land and property in trust for his children but would inherit it absolutely.

Next in the will came George's daughter: Samuel's wife, Mary-Ann. She would inherit the dividends and interest on £2,000 of stocks of 3 per cent consolidated annuities at the Bank of England. On her death the stock was to be divided between the Baxter children except twenty-two-year-old William. William would receive no stock but would inherit absolutely in his own right 10 acres of freehold marshland, also currently rented out. If William died before his grandmother and father, then the 10 acres would pass to Samuel.

There are some small bequests to more distant family members. George's deaf and dumb step-granddaughter Betsy Smith was to receive a not especially generous pension of £5 a year for life to be paid by the Baxters out of their share, while Elizabeth Evans and another of Ann Bodle's daughters by her previous marriage would receive £10 each, and old George's sister Elizabeth Andrews, £20. The executors Henry Mason and George Wassell were to have £10 each 'for the trouble they will have in the execution of this my will'. The payment was a gesture, hardly a recompense, particularly for Mason, for whom the responsibility would prove more troublesome than either he or George could possibly have foreseen.

But then the will comes back to the Baxters. Anything that was not already assigned – 'all the rest, residue and remainder of my goods, chattels, debts, credits and estate and effects and property of every description' – was to go to William and his heirs, unless he predeceased Ann, in which case it would go to Samuel.

William and Samuel were also to be executors along with Wassell and Mason. Middle John, therefore, was trusted neither with the outright possession of any property or investments, nor with any responsibility for ensuring that his father's instructions were carried out. George placed in the Baxters a confidence he apparently did not have in his own son.

There's Not Much in Dying

Henry Wallis's painting of the poet Thomas Chatterton has a sad finality about it. The seventeen-year-old poet died in a London attic in 1770 after committing suicide by taking arsenic. He was idolised as a tragic genius by the Romantics; Wordsworth called him 'the marvellous boy' while Keats dedicated *Endymion* to him. Wallis portrays the fully clothed body of a young man sprawled gracefully on a bed, his head with its tousled red locks turned towards the observer, his face and neck white as alabaster, the features serene. One arm rests across his breast, the other hangs artfully over the side of the bed.

Whatever its artistic merits, the work is not all it seems. The dead youth is not a likeness of Chatterton but of a young man called George Meredith, and the dignified scene Wallis portrays bears no resemblance to the aftermath of a death from arsenic poisoning. (There is a rather less tasteful theory about how Chatterton died – accidentally, from an overdose of an arsenic preparation used to treat a venereal disease.) If John Butler was right in his diagnosis of what was ailing George Bodle, then the old man was facing a dreadful end.

'There's not much in dying,' Emma Bovary tells herself after cramming her mouth full of white arsenic. 'I shall go to sleep, and it will all be over.' But Flaubert's heroine was appallingly mistaken, her imagination betraying her one final time: 'A violent shudder went through her shoulders, she turned whiter than the sheet she was clutching in her fingers ... Drops of sweat stood on her blue-veined face ... Her teeth chattered, her pupils were dilated ... A muffled scream broke from her ... "Oh God, it's ghastly!" she cried.' Soon she is writhing and vomiting blood. '... brown spots broke out over her body, her pulse slid between the fingers like a taut wire, like a harp-string about to snap. She began to scream horribly.'

To unleash its attack, arsenic has to be absorbed into the bloodstream so there is normally a delay of ten minutes or so – longer on a full stomach – before the victim starts to realise that all is not well. Then comes tightness and a burning sensation in the throat and gullet that makes it hard to swallow. As the poison enters the bloodstream, the blood vessels become inflamed, affecting the digestive system, causing stomach cramps and nausea, followed by violent, often projectile, vomiting.

Any food in the stomach is thrown off along with any arsenic that has not been absorbed, but instead of easing as the stomach empties, the retching continues relentlessly. By now the victim is vomiting blood and bile, perhaps even faeces pulled up from the gut by the violence of the retching. In between the bouts of vomiting come increasingly agonising stomach cramps, and then the purging begins as the body tries to replace the fluids lost from the stomach by draining water from other parts of the system into the gut, only for it to be expelled by the vicious contractions of the abdominal muscles. The initial watery diarrhoea, or what was described in the nineteenth century as 'rice-water evacuations', is then replaced by blood as the capillaries burst and the straining continues, even though the body has nothing left to expel.

If, after all this, the victim is still struggling for life, then he or she may experience a relief from the diarrhoea, only to find it replaced by painful involuntary anal spasms known as tenesmus – an urgent but fruitless desire to empty the bowels, usually resulting in an ulcerated anus. The effects of the poison then spread to other parts of the body, and the kidneys are affected, making urination painful, together with pain in the bones and muscles.

The huge loss of body fluid from all the vomiting and diarrhoea causes violent thirst but drinking is not only pointless but adds to the misery: any liquid swallowed is immediately thrown back up in another spasm of retching well before it stands any chance of being absorbed.

Some poisons such as cyanide and strychnine work according to a strict timetable and dispatch their victims in a predictable manner. Arsenic, by contrast, is mysterious and shilly-shallying, behaving more like an infectious disease, so that the nature of the victim's suffering and how long it will last depends partly on the individual's genetic make-up. And to confuse matters further, human beings are capable of building up a certain degree of tolerance to arsenic if they go about it carefully enough.

In 1851 Dr von Tschudi caused a sensation when he reported the case of Styrian peasants living on the borders of Austria and Hungary. They believed arsenic was good for them and took it as a tonic in what would normally be lethal doses. They started with half a grain two or three times a week and gradually increased the dose. (Lawyers quickly seized on this 'Styrian defence' to try to sow doubt in juries' minds. Was this really a case of murder? Perhaps the arsenic in the dead person's food or body had been self-administered for health reasons, only this time the victim had gone too far?)

Death from acute arsenic poisoning can take anything from two hours to four days, although victims have been known to linger for a fortnight. For most, though, the misery lasts at least twenty-four hours. But unlike cholera suffer-ers, arsenic victims do not die of dehydration: the strain on the heart causes a complete system collapse, ushering the individual towards the grave. To the victim, of course, the point is somewhat academic.

And if acute arsenic poisoning is not terrible enough, the slow and nasty decline produced by the sub-acute ver-sion, when the victim is fed small doses over a long period, is truly the stuff to frighten the horses. Loss of appetite, weakness, vomiting and diarrhoea are followed by weeks, or even months, of muscle pain, numbness and tingling in the hands and feet, urine retention and outbreaks of sores, scabs and blisters all over the body, ending with paralysis, convulsions, coma and death. Murder by sub-acute arsenic poisoning requires a special kind of determination and ruthlessness, which is why it was so hard to contemplate the idea that within a person's own household there could be someone who, while feigning love or loyalty, was gently scorching the life out of their victim.

An article in Charles Dickens' magazine *Household Words* declared that murder by slow poison admitted 'more readily of a fiendish sophistication in the mind of the perpetrator than any other form by which murder is committed'. The imagined perpetrator is a terrifyingly harmless-looking character: 'A thin, respectable-looking man in spectacles, with dark hair and whiskers and wearing a long brown coat, calls at a chemist's shop in a small country town one morning and asks for an ounce of arsenic to kill rats ... He has a design to poison his wife, her mother or a man to whom he owes money by small doses from time to time and he has now got a stock in trade for the carrying out of his intentions.'

Over the period from 1750 to 1914, arsenic in one form or another was involved in 237 cases to come before the English criminal courts. A long way behind – with fifty-two cases – came a plant-based alkaloid, a narcotic as readily available as arsenic, although in its pure form, more expensive at over 2s an ounce. People right across the social strata were addicted to opium, obtained from the dried juice of the poppy, *Papaver somniferum*. Opium was sold over the counter either in solid form as dried juice, or in a weaker, cheaper solution called laudanum, which was used as a painkiller, comforter and general panacea, and cost about the same as a pint of beer. No one was counting how many gallons of Godfrey's Cordial, a blend of opium, treacle and spices sometimes known as mother's friend or infant's bottle, were poured down infants' throats during the nineteenth century, but a few drops could kill a baby, and medical officers suspected that overdoses, as well as starvation due to a child being too doped to eat, were responsible for incalculable numbers of infant deaths.

In voicing his suspicions about George Bodle's death, John Butler had been careful to allude more generally to some kind of 'irritant'. In doing so, he was referring to the then standard method of classifying poisons as devised by Mathieu Orfila. Orfila saw off some of the myths that had been around for centuries, such as that the heart of a poison victim would not burn.

Born in Minorca in 1787, Orfila was a charismatic character, good looking and elegant, and in the 1820s and 1830s he was one of the most popular teachers at the Ecole de Médecine in Paris. His rich baritone voice allowed him to deliver with ease chatty, conversational-style lectures to over a thousand students at a time, and it was said that but for his fascination with chemistry, he could have made a fortune as an opera singer. Instead Orfila chose to devote his working life to the study of poisons, how to detect their presence, and their various antidotes and treatments.

From six in the morning until four in the afternoon, Orfila could be found at the school of medicine, coat off, shirtsleeves turned up to the elbows, his hair 'the sport of the winds', according to the Parisian gossip magazine *Charivari*, 'cutting, slashing, emptying, jugulating, broiling, boiling, frying, hurrying from one furnace to the other, stirring the coals, overlooking the broth and preparing nameless ragouts'. At four o'clock Orfila deserted his furnaces, washed his hands, assumed his neckerchief and sable coat and 'vanished like a dividend'.

In the pursuit of science, the toxicologist subjected hundreds of animals – mainly dogs and cats – to a dreadful death, and 400 people at a time paid 50 francs for the privilege of watching him. *Charivari* described the street outside the school on such occasions as resembling the Paris Opera on a much-anticipated first night. Inside, on a huge stage in the amphitheatre where Orfila normally gave his lectures, stood two furnaces emitting thick fumes, with a collection of 'queer-shaped instruments, unheard-of phials, tubes defying description' at the ready. In the background was 'a kind of small Roman altar on which is laid an unhappy dog'. Some of the sufferings to which these animals were subjected – such as the professor stepping on a dog's tail to prove to the audience that it was alive – seemed more like gratuitous cruelty than a regretful trade-off in the name of learning.

At least Orfila's victims did not die in vain. Systematically the professor went about describing the physical and chemical properties of scores of poisons and classified them according to their actions. He fed his subjects different amounts of different poisons, establishing the toxic dose for each substance and, as Benjamin Brodie had done earlier on a more limited scale, he recorded the symptoms produced and the time it took each animal to die. He would then dissect the body, taking detailed notes of the damage he found.

The result was the first comprehensive textbook on toxicology, published in 1814, and running into several editions. In the 1821 edition of *A General System of Toxicology* Orfila grouped poisons into four categories according to the symptoms they produced: the irritants (that is, those causing inflammation in the parts of the body they came into contact with); narcotics (affecting the nervous system or brain, producing stupor or delirium); the acrid-narcotics (believed capable of sometimes causing inflammation, sometimes stupor, sometimes both); and the septic or putrefactive poisons. This last group was a somewhat motley collection that included insect stings, bites from poisonous spiders and rabid dogs, eating poisonous fish such as dolphin and conger eel (presumably when bad), and also the condition called contagious malignant pustule, a form of anthrax caught by handling infected meat or animal skins and also known as wool-sorter's or rag-picker's disease.

The 1814 edition had also included 'contagious miasmata emanating from pestiferous bodies or bales of merchandise coming from a place infected with the plague; emanations from a confined space where a number of persons are shut up, receiving the air only through small apertures; exhalations from burying grounds, hospitals, prisons, ships, privies, marshes, putrid vegetables and stagnant water'. In other words, poisonous air.

The irritating poisons Orfila listed included both mercuric chloride and all the mercurial preparations; arsenic and arsenic compounds; verdigris and other salts of tin, gold and bismuth; silver nitrate, sal ammoniac (ammonium chloride), liver of sulphur, nitre (saltpetre); the salts of barites (barium salts), glass fragments, cantharides (Spanish fly, better known as an aphrodisiac); salts of lead; and what he called 'acrid plants or their concrete juices'. Corrosives such as sulphuric acid produced so much inflammation that they appeared to work in the same way as burns. Others were less caustic but were almost as quick to kill because they were absorbed into the bloodstream and carried around the body, destroying the heart, lungs, brain and nervous system, 'organs so essential to the preservation of the individual that death must be the inevitable result of an extensive injury received by them'. One of Orfila's chief breakthroughs came in 1839 when he extracted quantities of arsenic from the liver, spleen, kidneys, heart and muscles of a human corpse, showing that the poison was absorbed by all the major organs of the body, and not just confined to the digestive system.

Among the narcotics Orfila put opium, henbane (black nightshade), cyanide and yew. Under acrid-narcotics he included mushrooms, nux vomica, camphor, tobacco, belladonna, digitalis, rue, hemlock; 'emanations from flowers'; and what was known as horned or spurred rye. Horned rye is grain infected with the fungus *Claviceps purpurea*, which causes ergot poisoning in human beings. In its acute form, ergotism manifests as the hallucinations and convulsions known as St Anthony's Fire, while the chronic condition results in dry gangrene of the lower limbs.

On 'emanations from flowers' Orfila said: 'Persons who reside with impunity in rooms filled with odoriferous flowers have much difficulty in believing that some individuals are unable to remain even for a few minutes in such apartments without suffering symptoms such as headache and nausea, which with some are followed by convulsions and swooning.' Roses, lilies and honeysuckle were among the culprits, he believed. His recommended treatment was to leave the room at once and inhale vinegar fumes. If the patient did not recover fast then he should drink sugar and water.

Despite such anomalies as pestiferous bodies and floral emanations, Orfila's attempt to impose some scientific order on how toxins were defined was generally regarded as an important advance. The Scottish toxicologist Robert Christison, writing in 1836, said that until Orfila, the classification of poisons had 'hitherto defied the ingenuity of toxicologists'. Scientists had been content with grouping substances according to whether they were animal, vegetable or mineral.

Ten years younger than Orfila, Christison had attended some of Orfila's lectures in Paris when he was professor of medical jurisprudence (forensics). In 1822, the year after he returned from Paris, the twenty-five-year-old Christison was himself awarded a chair in medical jurisprudence, in his case at Edinburgh. He was said to have secured the job because his politics – firmly Tory – were deemed acceptable and because one of the decision-makers owed Christison's mentor a favour. However questionable the selection process, he proved to be a good choice. Ten years later he was made professor of materia medica, or pharmacology.

Christison's own treatise on poisons and their classification, published in 1829, was also a best-seller and went into four editions during the next fifteen years. Fish in particular troubled him, as they had Orfila. '... the oyster or muscle [sic], though generally salubrious and nutritive, will at times acquire properties that render them hurtful to all who eat them, and others, although eaten with perfect safety by mankind in general, are nevertheless poisonous either at all times or only occasionally to particular individuals', Christison wrote, adding: 'But hitherto the chemist and the physiologist have in vain attempted to discover the cause of their deleterious operation.' Not until the French chemist Louis Pasteur put forward his germ theory in 1858 did scientists begin to unravel the role that bacteria have to play in disease and decay.

In his classifications, Christison included what he called

mechanical irritants – fragments of steel and glass and fruit stones – describing a case where a woman was said to have died after two years of constant suffering after swallowing a large quantity of cherry stones. He did not explain how or why, but again Christison's mechanical irritants and Orfila's floral emanations showed the difficulty in arriving at a satisfactory definition of a poison; experts had struggled for centuries with the problem but everyone agreed with Paracelsus that it had something to do with the dose. Morphine was a good example. A quarter of a grain would kill a normal infant, but for a non-addicted adult in pain that dose would be therapeutic, while in an adult or even a child accustomed to the drug, it would have almost no effect.

More than three hundred years after Paracelsus, Alfred Swaine Taylor, another British toxicologist who studied under Orfila in Paris, pointed out that while the popular view of poison as a substance capable of destroying life even in small doses was useful enough for everyday purposes, it was inadequate as far as the courts were concerned. As the country's foremost forensic toxicologist at the time, Swaine Taylor was more concerned with how best to present the results of chemical analyses to judge and jury.

Like Christison, Swaine Taylor became a professor of medical jurisprudence at the young age of twenty-five, in Swaine Taylor's case at Guy's Hospital. He was to continue in this post until 1877, giving evidence at high-profile poisoning trials and becoming something of a household name. A tall, imposing man, he was 'gracious to friends and bitter to foes', according to the *Medical Times*, while the *British Medical Journal* saw him as 'Lucid in his exposition, conservative in his theories and elaborate in his investigations'.

One of the problems with defining a poison as something that was lethal in small doses was that it excluded chemicals that killed only in large doses. And 'In a medico-legal view, whether a man dies from the effects of an ounce of nitre [potassium nitrate] or two grains of arsenic, the responsibility of the person who criminally administers the substance is the same,' Taylor pointed out.

Epsom salts, or magnesium sulphate, commonly taken as a mild laxative or added to a warm bath for muscle pain, were not normally seen as harmful, but at Huntingdon Assizes in 1842 Thomas Stangon and William Saybridge stood trial for 'feloniously killing one Daniel Cox, by administering to and causing to be taken by him, a large quantity of Epsom salts mixed in beer'. Cox was an old man who spent his time wandering from pub to pub, cadging drinks and generally annoying people. One morning after putting away several half-pints at the Cross Keys in Sawtree, he began complaining that the beer was too weak. Two hours later he was seized by chronic diarrhoea and died the following night. A post-mortem report gave the cause of death as 'excessive purging and inflammation consequent thereupon'. Then the story came out: spiking Cox's drinks with substances such as tobacco and anything else that came to hand had become something of a local pastime, 'which the depraved palate of the old man did not enable him to discover', said *The Times*. But now the joke had gone too far. Saybridge was acquitted but Stangon was found guilty of manslaughter, although he was jailed only for a fortnight because he was said to have the 'highest possible character for humanity and goodness of disposition'.

Another problem, according to Swaine Taylor, came

about when ignorant lawyers played fast and loose with the term 'deadly'. 'Poisonous' or 'of a destructive nature' would do just as well, he thought, but the legal officers who drew up the indictments were 'but little informed on such matters and they can never speak of a poison without describing it as deadly'. He posited that the adjective held true only for those poisons that were lethal in very small doses and acted quickly, such as strychnine, cyanide and arsenic. Two people had been acquitted of attempted murder with copper sulphate because the indictment described the chemical as a 'deadly' poison and two medical witnesses could not agree on whether it was or not. Taylor's own definition in his first work on forensic medicine was: 'a substance which, when absorbed into the blood, is capable of seriously affecting health or of destroying life'.

The idea that a true poison had to be absorbed into the bloodstream gained ground over the century, leading people to wonder about hydrochloric, sulphuric and nitric acids. These were clearly fatal in small doses and they acted fast, but they spelled disaster from the first contact with the victim's skin and internal organs well before they entered the bloodstream. Did that mean, then, they should not be described as poisons? Some scientists argued that mineral acids could be both poisonous and corrosive, depending on how they were used. A teaspoonful of concentrated acid could kill within twenty-four hours, but a weak solution might have time to reach the bloodstream. The point was developed further with the example of a murderer pouring sulphuric acid into a sleeping victim's mouth as opposed to throwing him into a vat of the stuff. The action of the acid was the same in both cases, but the application was different; and whatever the first case was, the second was clearly not poisoning.

The law would eventually deal with the problem not by worrying away at the definition of a poison, but instead using the catch-all phrases 'stupefying or overpowering drug' and 'poison or other destructive or noxious thing'. This covered mechanical agents such as powdered glass; corrosives that destroyed the organs on contact; toxin-producing bacteria; and the 'true poisons' – those that entered the bloodstream and acted upon the blood, the internal organs or the deep-seated tissues. And while the scientists argued over the relative qualities of some substances, there was never any doubt about the status of arsenic.

6

A Great Degree of Inquietude

The Bodles were suffering horribly but at least they were spared leeches and the stomach pump. In the nineteenth century, bloodletting was one of the most common medical treatments. The ancient Greek system of medicine considered blood to be one of four 'humours', along with black bile, yellow bile and phlegm. Sickness was due to an imbalance in these bodily fluids and if the patient was to be brought back to health, harmony had to be restored. By 1833, the humours theory had long been abandoned, but the practice of applying leeches or opening a vein to release toxins was still widespread.

To treat arsenic poisoning, Orfila had recommended starting with twelve to fifteen leeches on the abdomen over the centre of the pain. If the pain moved, more leeches were to be applied to the new site. 'And even upon a third change of situation, we ought not to be afraid of applying 12 or 15 leeches more,' Orfila wrote, adding: 'The safety of the patient depends on the copious evacuation of blood; and the feebleness thus produced is to be considered as comparatively only a slight inconvenience.'

In 1824 England's first professor of forensic medicine, John Gordon Smith, wrote about a device that was being hailed for its success, although Smith said the idea was 'by no means a new thing'. 'The principle is that of injecting the stomach with warm water through an elastic tube, furnished at the external extremity with a syringe or gum-bottle, capable of containing about a quart,' he explained. 'The contents are then withdrawn by pulling up the poison on relaxing the bottle. The operation is to be repeated till the water comes off clear and tasteless.'

Other favourite remedies for poison cases at the time included milk, vinegar, linseed, copious amounts of sugar and water, and tickling the back of the patient's throat with the fingers or a feather. The aim was the same: to make the patient throw up the deadly matter as quickly and copiously as possible.

There had been some attempts to find an antidote to arsenic over the years and the remedies being touted at the time included iron perchloride, itself highly toxic, and iron oxide, which, it was thought, would bond with the white arsenic to form an insoluble salt that would then pass safely out of the body. The Covent Garden chemist Joseph Hume was said to have had some success with magnesia, while others recommended powdered carbon.

In 1809, the Devonshire physician Bartholomew Parr thought it worth trying a scruple (20 grains or one third of a gram) of vitriol (sulphuric acid) to induce vomiting, followed by large quantities of olive oil, gum arabica, milk and fatty broth in order to 'glue' together any remaining traces of poison and allow them to be thrown up in one mass.

George Male, a physician at Birmingham general hospital, recommended potassium carbonate or, failing that, sulphur, wood ashes or soap dissolved in a solution of quicklime. Male died in 1845 from aconite or monkshood poisoning, self-administered after he read that it was an effective treatment for rheumatism.

When John Butler had finished examining his patients that Saturday night, however, he prescribed nothing more harmful than egg whites and castor oil, both standard treatments at the time. And despite his confidence in leeches, Orfila also believed this method, an emetic of egg whites in cold water, still to be the best remedy. The typical practice was to administer twelve to fifteen eggs beaten in two pints of water every two or three minutes. The Bodle women followed Butler's orders but old George was having none of it. Probably wanting to assuage a raging thirst and burning throat, he ignored the doctor's unpalatable advice and drank a pint of ale instead, which Butler thought disastrous. As far as the old man was concerned, Butler felt he was fighting a losing battle.

Early the next day, Sunday, 3 November, the surgeon was back at the Plumstead farmhouse. Four of his patients, although still very sick and weak, were showing signs of improvement, particularly the younger women, Betsy and Sophia. The farmer, however, was much worse. When his daughter Mary-Ann Baxter and her husband Samuel arrived as usual from North Cray for Sunday dinner they were horrified at what they found.

Butler now decided that he was out of his depth with George Bodle; the time had come for a second opinion. For him, the obvious man to consult was the veteran physician

Dr Sutton of nearby Greenwich, a former army doctor and consultant to the charity the Kent Dispensary.

In 1833 any man could describe himself as a doctor and might hold any one of a bewilderingly wide range of qualifications. At worst, he might have no qualifications at all, nor have undergone even the most rudimentary training. In 1834 the president of the Royal College of Surgeons, George Guthrie, imagined the case of a young man who had been working as a druggist's assistant for four years. The man would then complete the two years of study required by the Company of Apothecaries and the Royal College. 'He probably had no preliminary education, in some instances he may be little more than able to read and write, and as to the dead languages [Latin and Greek] ... he knows nothing about the matter.' If this man then failed to qualify as an apothecary or a surgeon, 'he goes back to his native town, sets up a shop and practises as a chemist, druggist, surgeon, apothecary and man-midwife'. Under the Apothecaries Act of 1815, he could be made to take down any sign describing himself as an apothecary, should the Apothecaries find out, 'but the College of Surgeons has no such power'.

The Apothecaries Act had been the first step in regulating the profession, setting standards for training before students were allowed to sit the licence exam, but there was still a long way to go. The Royal College of Surgeons' qualification 'confers no legal privilege on its possessor, while any individual in the community is as much entitled to engage in the practice of surgery as any person who has had the folly to pay £22 guineas [the college's exam fee] for a piece of disfigured parchment', *The Lancet* declared, adding that the possessor of the Society of Apothecaries' £6 licence was superior to the surgeon and might laugh at his pretensions.

Not that *The Lancet* was particularly impressed by the Worshipful Company of Apothecaries, describing the law that required anyone sitting the licentiate exam to have served an apprenticeship as being 'conceived in the true spirit of a corporation of shopkeepers'. A medical student was compelled to spend five of the most valuable years of his life 'in a state of vassalage and ignorance behind a counter'. And such was the lack of rigour of the exam itself, said the journal, that beside it the farcical goings-on in Molière's medical satire *Le Médecin Malgré Lui* 'ceased to be a caricature'.

The Lancet's jibes about shopkeepers and counters hit a tender if well-worn mark. The apothecaries were considered to be on the lowest rung of the medical profession's three-tier hierarchy and had been trying since the sixteenth century to shake off their association with grocers. Rebranding had proved something of a struggle, particularly with people such as *Lancet* editor Thomas Wakley around to remind them of their roots. An apotheca was a store selling wine, herbs and spices, hence anyone selling these items was an apothecary. In the fifteenth century, when the guilds of pepperers and spicers came together, the Worshipful Company of Grocers was set up, taking in the apothecaries. From here a specialist branch, the spicer-apothecaries, emerged in a part of the City of London near Blackfriars called Bucklersbury, where they sold their spices, sweets, perfumes, spiced wines, herbs and drugs.

By the mid-sixteenth century the spicer-apothecaries had effectively become community pharmacists, preparing and selling medicines. They began their campaign to break away from the grocers and, in 1617, they succeeded and the Worshipful Society of Apothecaries of London received its royal charter.

But while the apothecaries prepared medicines, the physicians, the elite of the medical profession, controlled the prescribing. At that time apothecaries, unlike physicians, were not supposed to charge fees for giving medical advice but to earn their living solely by charging for the medicines they supplied. This rule was frequently broken, however, and the two groups jostled for power and control. Thomas Cook's comedy *Physick Lies A-bleeding, or the Apothecary Turned Doctor* features Tom Gallypot, 'an apothecary by trade, but who practises physic as a doctor near Covent Garden'; Lancet Pestle, 'an apothecary by profession (but who boldly undertakes to be a physician and surgeon)'; Retorto Spatula d'Ulceroso, 'an apothecary in Drury Lane, who pretends to be a great doctor, surgeon and chemist'; and Jack Comprehensive, 'an apothecary living in Fleet Street, who professes himself merely to be a doctor, surgeon, chemist, druggist, distiller, confectioner and (on occasion) corn-cutter'. The play is subtitled: 'A comedy, acted every day in most Apothecaries Shops in London, and more especially to be seen by those who are willing to be cheated, the first of April, every year. Absolutely necessary for all persons that are sick, or may be sick.'

Next up the ladder from the apothecaries came the surgeons, who had an embarrassing ancestry of their own. In the mid-fifteenth century the then Fellowship of Surgeons amalgamated with the Company of Barbers to form the Company of Barber-Surgeons, carrying out minor procedures such as bleeding, cupping and pulling teeth as well as cutting hair and shaving. (The traditional red-and-white-striped barber's pole symbolises blood and bandages.) But as interest in the study of anatomy grew, surgery, though still a crude and bloody business, developed into a specialised area and the surgeons moved to reclaim their independence. The new Company of Surgeons' first hall stood close to the Old Bailey and Newgate prison, which was handy for collecting the corpses of hanged criminals, the only legitimate source of bodies for dissection at the time. By 1800, however, the surgeons had been granted a royal charter and they moved to a purpose-built home in Lincoln's Inn with lecture theatres, meeting rooms and space for the anatomist John Hunter's magnificent collection of anatomical specimens.

At the top were the physicians, who could trace the origins of their profession back to the ancient Greeks. From the time Henry VIII set up the College of Physicians in 1518, the institution was engaged in a power struggle with the other medical bodies. The College wanted to be able to grant licences to those qualified to practise medicine and punish those who were not. Their charter decreed that they would: 'curb the audacity of those wicked men who shall profess medicine more for the sake of their avarice than from the assurance of any good conscience, whereby many inconveniences may ensue to the rude and credulous populace'.

The College of Physicians was set up to be an academic body rather than a trade guild – not for them the drawing of teeth in the marketplace or the selling of herbal remedies from a stall. They were the elite of the medical world. At first candidates for the College licence needed an Oxford or Cambridge degree to prove they were 'groundedly learned', in other words to pass an oral exam in Latin and Greek. The College finally agreed to admit licentiates from non-Oxbridge universities in 1835 but Fellows still had to be Oxbridge men.

John Butler had followed a respectable path into medicine. By the time he sat the Society of Apothecaries' exam he was overqualified, with six courses on anatomy, six on the theory and practice of medicine, six on chemistry and four on materia medica (pharmacology) behind him. He had also spent six months at Guy's Hospital, near London Bridge, walking the wards. He passed the exams both for the Society of Apothecaries' licence and membership of the Royal College of Surgeons. This standard dual qualification was known as 'College and Hall', the latter a reference to Apothecaries' Hall where the examinations were held.

While doctors such as John Butler were described as surgeons because of their membership of the Royal College, only those at the top of the profession – such as the famous Sir Astley Cooper, who had removed a cyst from George IV's head and taught John Keats when the poet was a medical student – were able to earn a living from surgery alone. Most surgeons in towns and villages across the country were really general practitioners – prescribing medicine, performing operations, dressing wounds, setting broken bones, letting blood, pulling teeth, applying 'blisters' or poultices, and delivering babies. The term 'general practitioner' was unknown in the eighteenth century but by the 1830s it was more commonly recognised.

*

The sixty-six-year-old Thomas Sutton came with a considerable reputation. He had studied medicine in London, Edinburgh and Leiden, and with his medical degree and his membership of the elite Royal College of Physicians, he was at the top of the medical hierarchy. An intelligent and thoughtful man, Sutton had published papers on a wide range of conditions, including tuberculosis, peritonitis and gout, but he was best known for his work on delirium tremens or the 'drunkard's madness'.

Butler's choice of Dr Sutton was based not so much on his academic standing, however, but on personal experience, for the links between the Butler family practice and the Greenwich physician went back at least twenty years.

In 1813, John Butler's father was treating a woman with puerperal fever. Puerperal, or childbirth, fever is the result of an infection in the womb or vagina, which typically strikes a few days after a woman has given birth. Before the need for strict hygiene was understood, this was a major killer, and tragically most cases were caused by the doctors themselves not washing their hands before examining women, sometimes going straight from an operation or even a post-mortem to deliver a baby.

When John Butler's father sent for Sutton in 1813, he had just lost three patients to the condition and now, despite his best efforts, a fourth seemed set to follow. In each case the woman had had a straightforward labour and seemed well for the first two or three days. When Sutton saw Butler's latest patient six days after delivery, however, she was pale, feverish and exhausted. Sutton's remedies for the patient included the traditional large doses of opium for the pain, but he also prescribed a less orthodox treatment:

the application of a 'cold' lotion to the painful parts in order to reduce the inflammation.

When Sutton next ran into Butler senior he was delighted to hear not only that the patient had survived but other women too appeared to have been saved by this method. Sutton and Butler senior also met through their charity work at the Kent Dispensary, which provided free medicine and medical care to the poor of Woolwich, Deptford and Lewisham.

On Monday, 4 November, then, Thomas Sutton arrived at Bodle's farm, where John Butler went through the background to the case and described George Bodle's symptoms. Unlike the Butlers, who as provincial surgeons were used to getting blood on their hands as well as their arms and aprons, physicians did not examine patients nor administer treatment but asked questions, listened to descriptions of symptoms and then arrived at a diagnosis. Any physical contact was usually restricted to taking the patient's pulse.

After the briefing, Sutton proceeded to the bedside to cast an eye over the suffering old man. He noted that while the farmer had more or less stopped vomiting and was no longer complaining about his eyesight, his pulse was slightly high at 120 and he still professed to a terrible burning pain in his stomach that became worse when pressure was applied. Bodle was, Dr Sutton said, 'labouring under a great degree of inquietude', and he gave his expert opinion that Butler was right: 'some deleterious matter had been administered'. Unfortunately in this case Sutton had no innovative treatment to suggest: his younger colleague was already doing all that could be done, medically speaking. Hoping and praying were all that was left.

As to how the deleterious matter had found its way into the patient's system, Sutton was not only incurious but seemed surprised that anyone should think he might be interested. 'It was not necessary to enter into that question,' he later explained. 'I had received information from Mr Butler of the symptoms and I did not conceive it my duty to make inquiry into the circumstances.' Nor did he consider it within his remit to ask George Bodle if he thought he might have been poisoned. When asked whether the farmer believed he might be dying, Sutton replied: 'That was not a part of my professional inquiry. I had made inquiries of the symptoms before I saw him.'

Clinically Sutton was right, of course: how much George knew about his situation and its likely end was of no use whatsoever to a physician as he tried to arrive at a diagnosis and decide on treatment. Legally, though, in a case of suspected murder, it could be vital. While hearsay evidence is not usually admissible in court, an exception can sometimes be made, for example if a dying person's statement relates to the manner or cause of his death. But there is an important proviso: when the statement is made, the victim must be under no illusion that he has the slightest chance of recovery. For he would not risk his soul by dying with a lie on his lips, or so the thinking went in the nineteenth century. And who better to know the state of mind and hear the confidences of a dying man than the doctor attending him?

In 1789, when William Woodcock stood trial for battering in the skull of his wife Silvia, Sir Robert Eyre, Lord Chief Baron of His Majesty's Court of Exchequer, stressed the importance of dying declarations being made 'in extremity, when the party is considered as at the point of death, no hopes for this world, with every obligation upon them therefore to speak truth'.

A watchman had found Silvia lying in the street 'very much beat and bloody', with one of her ears 'cut or beat off' (it was later produced in court). He had taken her to the nearby Chelsea workhouse where she was seen by the surgeon Thomas Manby Read and the apothecary John Powell. Over the next few days, as Read and Powell dressed her wounds, Silvia recounted how her husband had 'beat her about her head and face in a most horrid manner'. At first the doctors had hoped to save her but when Read saw 'the putrefaction coming on very fast', he realised that they were losing the battle. At William's trial, though, the men were unable to swear that Silvia had known for certain that she was dying, although they were sure she knew she was very ill. In the end it mattered little; there was a mountain of other evidence against Woodcock, and five days later he was hanged.

Throughout the day after Dr Sutton's visit, the rest of the Bodle household continued to recover, but George's condition grew steadily worse. His son-in-law Samuel Baxter spent most of the day at his bedside. By the next morning, Tuesday, 5 November, four days after George had first fallen ill, it was clear that the farmer had not much longer to live. Mrs Lear stayed at her nursing post, where she was joined in the afternoon by another household help and relative of Ann Bodle, the sixty-year-old Ann Wooding. The longest the faithful Judith Lear left her master was for five minutes to fetch candles as dusk fell, while Ann Wooding turned away only occasionally to warm herself by the fire, for the room was cold. Mrs Lear found it 'a trying scene' to sit watching over the poor old man, now in no doubt that life was ebbing out of him, and she listened as he prayed fervently to the God he had served all his life.

He was still saying his prayers when his wife, very sick and frail herself, crept into the bedroom to say goodbye. He said 'You are come' and seemed pleased, Mrs Lear said. He had no strength to keep up a conversation but the couple prayed together.

Middle John had also been at the farmhouse every day since the Saturday evening when, looking in to pick up his wages as usual, he had found his father in the kitchen, sick and alone. He took over from Samuel Baxter at his father's bedside on Tuesday, and it was while he was huddled next to George in the dim light of a late autumn afternoon that, according to him, the old man had whispered into his ear a secret – confessing his suspicions as to how his righteous life had been brought to such a sad and painful end. For the time being, Middle John told no one what passed between them and he left the farmhouse at around four to go back to his cottage for tea. No sooner had he reached home, how-ever, than a servant came running down the path, saying that the old man was near his end. Middle John hurried back. 'I entered the room before he was quite gone, for I put my hand under his chin to prevent it falling,' he said. 'My father kissed me and said "God bless you" in a very faint voice.'

Mrs Lear, still keeping her bedside vigil, was glad to see father and son on such good terms, for it had not always been so. Indeed, she had been confused when her employer had turned to her earlier that morning and demanded, 'Has that fellow been here since? If he comes down again, turn him out of the kitchen.'

'Who do you mean?' Mrs Lear had asked.

'John Bodle.'

'Do you mean your son?'

'No, Young John Bodle. Let Henry Perks or someone else come for the milk or they shan't have it.'

If John Butler also suspected a particular individual of being behind the events at the farm, the surgeon was keeping his counsel. But he clearly felt this was no accident. That very night, 5 November, as church bells rang out across the county and young people took to the streets waving effigies of 'Guy Faux' and 'old Pope' on swords and hop poles, the surgeon was called to the house again. This time, though, it was to examine the racked body of an old man and to confirm that life was indeed extinct. This last duty done, the doctor left the weeping Mrs Lear and Mrs Wooding as they laid out their master's corpse and the rest of the family grieved, and headed his phaeton back towards Woolwich in search of a magistrate. The healers of the sick could do no more. Now it was the turn of His Majesty's criminal justice system.

7

Corroborative Proof as to the Deleterious Article

The entire success of a criminal poisoning depends on the process imitating the effects of a natural disease. So said Edinburgh's professor of forensics, Robert Christison. If he was right, then in 1833 the odds ran strongly in the murderer's favour because so many diseases were poorly understood.

In 1829 two Guy's Hospital men, pharmacology lecturer Thomas Addison and surgeon John Morgan, compared the symptoms of tetanus (a bacterial infection) with those of strychnine poisoning and found what they described as 'the most perfect resemblance'. 'In both cases we find the same spasmodic contractions of the voluntary muscles ... and in both cases the morbid action commences at the same point, viz. in the muscles of the lower jaw.' If diagnosing most types of poisoning was difficult enough, arsenic was especially tricky because it mimicked so closely the symptoms of several common diseases.

Until 1831, there were four conditions thought most likely to lead the medical man astray in cases of arsenic poisoning: English cholera, ataxique fever, the bloody flux and 'vomiting of black matter'. As mentioned, English cholera was a blanket term for the various viruses and bacteria that affect the digestive system, producing vomiting and diarrhoea. With food hygiene poor and living conditions often appalling, English cholera was regarded with a familiarity almost bordering on contempt unless the attack was particularly severe.

What was known as ataxique or remittent fever was probably a form of malaria, then prevalent in Britain in areas close to low-lying marshes and stagnant water. The symptoms included fever, jaundice and vomiting green bile. While no one knew at the time that malaria was carried by mosquitoes, by the 1830s doctors were well aware of the curative powers of quinine, known as Peruvian bark.

The bloody flux, or simply flux, was dysentery, a potentially fatal inflammatory condition of the intestine that causes severe diarrhoea with mucus or blood in the faeces, fever and abdominal pain. Vomiting of 'black matter' probably referred to the sufferer bringing up blood that had been acted on by the gastric juices. Such matter had the appearance of coffee grounds and usually indicated a gastric bleed.

This was all deeply confusing at a time when doctors had nothing to go on but the symptoms and the circumstances leading to those symptoms. Then, in the autumn of 1831, a disease appeared in Sunderland, north-east England, the like of which had never been seen before in Britain, although its arrival was hardly a surprise. For five years Asiatic cholera – or Indian cholera, epidemic cholera or *Cholera morbus* as it was variously known – had been making its way steadily out of India, north-west through the Russian Empire and across Europe, until it reached Hamburg on Germany's Baltic coast. From there it was but a thirty-six-hour crossing over the North Sea to some of England's busiest seaports.

In its severest form, cholera is one of the fastest killer diseases known to man, and a victim can die within two hours of the onset of symptoms. The government had tried to prepare for its arrival but was at something of a disadvantage: no one understood how the disease was transmitted, so no one knew how to prevent it spreading. (In fact, it is spread mainly through contaminated water but this idea was unheard of at the time.) And while there were countless bizarre and sometimes fatal remedies on offer – including slapping hot poultices smeared with irritants such as mustard on to patients' stomachs, pouring boiling water over their feet, giving them draughts of medicine containing mercury, ammonia or turpentine, and, of course, bleeding – nothing seemed to work. From Sunderland the disease fanned out across the country. Within twelve months 32,000 people had died. The symptoms – violent vomiting and diarrhoea, agonising cramps, skin turning dark blue or black, and a rapid decline – did mimic those of the severest cases of English cholera, and of arsenic poisoning, although army doctors in India who had seen it at close quarters said that *Cholera morbus* was unmistakable.

The Lancet predicted more trouble for the criminal justice system. In poisoning trials the defence often claimed that the cause of death was not arsenic at all but 'the ordinary species of cholera'. In trying to firm up a diagnosis, the medical profession had tended to use the rule of thumb that patients with English cholera as opposed to arsenic poisoning had no blood in their stools or vomit and that the disease was never fatal in less than twelve hours. Asiatic cholera rendered this distinction useless.

Nor was the circumstantial evidence quite so helpful any more. Robert Christison had told doctors to be on their guard if several people fell ill at once soon after a meal – especially if one member of the family was on bad terms with the rest – but Asiatic cholera defied this notion too. *The Lancet* reported a case where a family of four had sat down to dinner in apparent good health yet within a few hours three were vomiting and suffering from diarrhoea. The father's stools were tinged with blood and seven hours later he was dead. The mother appeared to be less affected than her husband but died four days later. The daughter's symptoms were initially more violent but she recovered the next day. The son, though, described as the 'mauvais sujet' of the family, suffered no ill effects whatsoever. 'Strong suspicions were excited but a minute investigation led to such ample proofs of the operation of the epidemic that these suspicions were completely allayed,' *The Lancet* reported. The journal did not explain what these ample proofs were but they probably saved the 'mauvais sujet' from the hangman.

John Butler's night call at the residence of the Reverend Dr Watson, JP, marked the start of the criminal investigation into how George Bodle had died. In 1833 the usual way to trigger an inquiry into a suspicious death was through the family or a doctor informing a magistrate or a coroner. Sometimes servants' gossip was the conduit, trickling through the neighbourhood until officialdom began to take notice. If a magistrate thought the matter sufficiently worrying he could order a constable to search the premises in question, bring in witnesses or make an arrest, but the magistrate, not the police, directed the investigation and questioned the witnesses and suspects.

Samuel Watson had known George Bodle well: they were both leading players on the limited local stage. Now when the magistrate heard what John Butler had to say, he did indeed think the matter sufficiently worrying and he determined to inform the coroner. Then, just after daybreak the following morning, Wednesday, 6 November, the magistrate received more unexpected callers who came to him regarding the same business. In the light of what they told him, he summoned the constable and issued further instructions.

In the meantime John Butler returned to the farmhouse to continue with some investigations of his own. It would be another nine years before the Metropolitan Police's detective department, the forerunner to the CID, was set up. In 1833 it fell by default to the first medical man on the scene to seize anything that might be considered useful evidence and to note any suspicious circumstances. Whether that medical man was up to the job or had any interest in performing it was another matter. Most apothecary-surgeons knew nothing about forensics, and even if they had known what to do, many would have regarded tasks such as scraping flakes of dry vomit from a sickroom floor as being beyond their professional remit.

When George died, the rules about what was accepted as evidence in crimes against the person had started to change, and nowhere was this more evident than in cases of poisoning. In fact, the story of forensic science in the first half of the nineteenth century is largely the story of forensic toxicology. In the eighteenth century, the guilt or innocence of the person in the dock rested mainly on circumstantial evidence and the character and demeanour of the accused; what was known as the moral evidence. Any medical evidence usually consisted of the parish surgeon or the doctor who had been called giving their opinion as to the cause of death and their interpretation of the symptoms. This was sometimes, but by no means always, coupled with the results of a post-mortem. Of 158 murder trials at the Old Bailey from 1829 to 1838, fewer than half – seventy-four – had both medical testimony and a post-mortem report to guide them.

Increasingly, though, as science advanced in the late eighteenth and early nineteenth centuries, the authorities began to appreciate that matter such as the victim's stomach contents, vomit and leftover food, drink and medicine had a tale to tell, and to interpret the evidence they relied on both doctors and analytical chemists. For the doctors' part, as their knowledge grew they were drawn into a better understanding of specialist areas such as dissection, pathological anatomy and analytical chemistry – all key planks of forensic medicine. They also began to see how they might use these new skills to help solve crime and at the same time improve the status of their profession.

The Bodle investigation was fortunate in that John Butler realised he had a key part to play in uncovering what he was increasingly convinced was a murder. He was dealing not only with a sickroom but a crime scene. Dr Sutton, on the other hand, showed not the slightest inclination to play detective.

In 1825, when John Butler had obtained his Society of Apothecaries licence, forensic medicine, or medical jurisprudence as it was usually known, was in its infancy in Britain – unlike on the Continent, where the subject had been recognised for decades. Lectures in forensics began at Edinburgh University in the late eighteenth century and a chair was established in 1807, but when Robert Christison became professor in 1822, the subject was still not mandatory.

In 1828 the new University of London created England's first chair in medical jurisprudence and appointed another Scot, John Gordon Smith, as its first incumbent. At first the medical school offered no degrees but merely coached students for the Company of Apothecaries' licence and membership of the Royal College of Surgeons. Gordon Smith was five years older than Robert Christison and, like the toxicologist, had qualified as a doctor at the University of Edinburgh. After serving as an army surgeon during the Peninsular Wars and at Waterloo, he had settled in London where he wrote *The Principles of Forensic Medicine*, one of the first comprehensive books on the subject in the English language. Two years later he was teaching forensics at the private Webb Street medical school and went on to lecture at the Royal Institution, alternating with the famous Michael Faraday's sessions on chemistry.

Unfortunately Gordon Smith got off to a bad start as a professor and he never really recovered. When he complained about not having enough time to cover all his material, the *Morning Chronicle* commented: 'In this regret we could scarcely join …. Condensation, we are afraid, is not a virtue of Dr Smith's.'

But Gordon Smith's real problem lay in the fact that his subject was optional, which meant that hardly anyone bothered to turn up, hitting his pocket as well as his ego because most of his salary came from student fees. So a year into his tenure he petitioned the Home Secretary, Sir Robert Peel, for medical jurisprudence to be made compulsory. Any doctor might be called upon to help investigate a crime, he argued, but while doctors could be trusted in the sick chamber, their contributions to public inquiries were 'a matter of great dissatisfaction to all the legal authorities of the kingdom'. In fact, Gordon Smith claimed, the notorious Burke and Hare case in Edinburgh just months earlier – where up to thirty people had been killed and their bodies sold to medical schools – would not have been possible if only the customers, the teachers of anatomy, had been trained in forensics and were therefore able to tell a victim of murder when they saw one.

Robert Christison might have argued with Gordon Smith that the line of demarcation between a natural death and a criminal offence was not always so easily drawn, however. He had examined the bodies of some of Burke and Hare's victims and had struggled to establish how Margaret Campbell, for one, had died and whether her death was an accident or murder. She was found with her head forced down on to her chest, a mark around her neck, a spinal injury, multiple bruises, a 'livid' face and lips, and with heavily bloodshot eyes. Strangulation was a possibility, Christison thought, but he was troubled by the spinal injury. Did it occur before or after death? If before, it must have happened during her final twenty-three hours, for she had been seen very much alive and dancing until then. Finally he had come to some sort of conclusion: 'I gave my opinion that death by violence was very probable.'

In September 1830, Gordon Smith won his fight and the Society of Apothecaries announced that in future all those studying for its licence would have to attend a three-month course in medical jurisprudence. The Company of Surgeons and College of Physicians followed suit. Sadly, though, for Gordon Smith, just as forensic medicine was gaining the recognition he craved, he himself fell out of favour. Ironically for a trail-blazer, he was seen as outdated, his lectures anecdotal and lacking the disciplined, scientific approach that the courts were starting to demand. The crunch came when Smith spotted some of his own books, personally donated to the university, on the library rubbish tip. He resigned the same year that his subject became mandatory and the faculty accepted his resignation with what appeared to be relief.

Beset by resentment and alcoholism, Gordon Smith died in the Fleet debtors' prison in 1831 aged forty-one. At the inquest, the coroner for the City of London recorded that his death had come about 'by the visitation of God', a common verdict at the time. In earlier, more literal days, it implied that the individual had in some way incurred the wrath of the Almighty, but by the nineteenth century it was simply an impressive way of saying that no one – on earth at least – had a clue why the person had died.

Medical jurisprudence was still optional when John Butler qualified, but even so he appeared to be following the instructions of specialists such as Gordon Smith and Christison to the letter.

In his guidelines for the detective-doctor, Christison advised looking for any suspicious behaviour before the fact: someone dabbling with poisons, discussing them at length 'or otherwise showing a knowledge of their properties unusual in his sphere of life'. If a suspect claimed he had administered the poison by mistake or didn't know what it was or wrongly thought it would be beneficial to the victim, the doctor should ask himself whether this was credible. And while the victim lay ill, did anyone try to prevent a doctor being called? Was anyone anxious not to leave the victim alone with another person? Did they try to dispose of any food, drink or vomit? And after the death, did they try to speed up the funeral arrangements or prevent people from inspecting the body? Then of course there was motive: had the suspect quarrelled with the deceased? Did the suspect expect to inherit? Was the deceased having his child and, if so, did he know about it?

So when Butler had first arrived at the farm that Saturday night he had set about questioning the family. As the Bodles told their tale and the surgeon's suspicions grew, he had begun to write down their statements in their own words. The surgeon then turned his attention to the scientific evidence. At that point the old farmer was still being sick so Judith Lear was instructed to save a specimen of George's vomit. In cases where the patient had stopped vomiting by the time the doctor arrived – perhaps because the patient was now dead – Swaine Taylor had advised that the sickroom be searched for stains. Any vomit on clothes, upholstery or carpets should be cut out of the cloth; any matter spilled on a wooden or stone floor should be scraped up or collected with a rag soaked in distilled water.

And the doctor should also keep an eye out for any suspicious-looking bottles or packets. If the patient was still alive then a small amount of vomit should be tested quickly by any means at hand in order to identify the poison and allow the correct treatment to begin, but the bulk of the sample was to be tightly sealed in a clean container and set aside for more careful analysis 'in order to speak to the purpose in a court of justice', said Gordon Smith. And if a recognised antidote worked, then this was not only a successful clinical outcome but further confirmation, should it be needed in court, that the doctor had correctly identified the poison – 'a corroborative proof as to the particular deleterious article'. Unfortunately for George Bodle there was no recognised antidote to the poison that John Butler suspected.

Finally, Smith stated, the doctor should ask himself 'whether any mistake or deleterious interference could have occurred in the culinary department'. So when Butler learnt of the ritual of the coffee pot, and with deleterious interference firmly in mind, he sent Judith Lear hotfoot down the road to her daughter Mary Bing's house to see whether by chance any trace of that coffee remained. He was in luck. Earlier that day, as the Bodle household collapsed around her, Mrs Lear had sent an urgent message to the Bings not to touch the jug she had left on their kitchen table. By the time the news arrived Mary's eldest daughter, twelve-year-old Eliza, had already added more water to the dregs, boiled up what by that time could hardly have been an appetising brew and poured herself a cup. Fortunately at the last minute the child had decided that the mixture looked thick and strange, and had tipped it back untasted into the pot. Her grandmother found it that evening on the mantelshelf, took it back to the farmhouse and gave it to the surgeon.

Butler then asked about the vessel used to boil the water for George and Ann's coffee and young Henry Perks was told to retrieve the scrapings from the garden where he had thrown them after George had ordered the kettle to be scrubbed. Next on the surgeon's list was the dry coffee that was kept locked in the parlour cupboard. By then the bottle was two-thirds full because after Sophia had used the last of the coffee there for breakfast it had been refilled. 'I therefore inverted the bottle and allowed the coffee gradually to run out and I found adhering to the bottom a coating of old coffee about one-eighth of an inch thick. This was evidently old coffee and of a different colour to that which I had turned out,' Butler explained. 'I could not say how long that coating had been there. There was about a large teaspoonful.'

He removed what he could of the coffee crust and took it away with him, along with the coffee pot recovered from the Bings' cottage, the fur from the kettle and the slop of George's vomit that Mrs Lear had collected. 'The suspected substance when once placed in his hands should never be left out of his sight or custody,' Alfred Swaine Taylor had warned after a judge in Norfolk refused to hear evidence about arsenic in a victim's vomit. The surgeon had given the specimens over to the care of 'two ignorant women' who had left them lying unattended as other people came and went.

After all of the unfortunate John Gordon Smith's struggles, once medical jurisprudence was recognised, it moved quickly up the agenda. In 1833, two years after Smith died, the University of London's professor of comparative anatomy, Robert Grant, another Edinburgh University man, gave the address marking the start of the medical school's academic year. He described medical jurisprudence as the highest branch of the curriculum, the most responsible department of the profession and the one requiring the most extensive and varied attainments.

'How many a victim has sunk unheeded into the grave and left the murderer to steal unsuspected through the rest of life?' said Grant. 'But now the carcass, gone to shreds, can be made to stand in appalling judgement against the murderer, and render him the poison, grain for grain.' The rhetoric was wonderful; the reality turned out to be less so.

8

Those Low Incompetent Persons

With George's death, the stories already circulating in the village of Plumstead, and especially in the Plume of Feathers, the Prince of Orange and the Green Man, reached new and delicious heights. In this small community, news travelled fast.

The fields, orchards, market gardens and scattered small-holdings of the old parish of Plumstead and East Wickham lay alongside the fertile marshland that bordered the River Thames to the north, while to the south the ground rose sharply to Plumstead Common, Shooters Hill and the Dover Road. At over four hundred feet, Shooters Hill is one of the highest points around London, with a dangerous past, as the name implies, its steep, dark woodlands providing high-waymen with both cover and a vantage point.

Little Plumstead, however, was untainted by this racy reputation. Though only ten miles from London Bridge, at the time of George Bodle's death it remained isolated and largely untouched by the rapid industrialisation that was transforming much of the English landscape. This was soon to change with the coming of the railways and a building boom. Even by the 1830s when George died, the signs were there: the population had more than doubled since the turn of the century, from 1,200 in 1801 to 2,800 in 1831, and overcrowding was on the increase. Still, for the time being life in Plumstead continued to follow the pattern of the seasons, the demands of land and livestock and the calendar of the Church of England, as it had for centuries.

The village itself consisted mainly of rows of cottages, a few large houses belonging to the likes of George Bodle and four pubs strung out along the high street, with a few smallholdings scattered beyond. The smallest cottages, consisting of just two rooms and fetching rents of about £3 a year, came with an apron of land a few yards square where the occupants grew their vegetables. Families such as the Bings, the Febrings and the Abletts scraped along with only parish relief between them and destitution. The men and the older children were hired by the week or the day and laid off when the weather was bad. Winter was something to dread. Harvest was when they made the most money and whole families often worked side by side, the women bringing their babies into the fields wrapped in a blanket or a cloak. Women also did some of the lighter work, pulling turnips, tying the young hop shoots to poles and picking fruit. By 1833, though, the economics of agriculture had changed considerably for the worse, with farm workers struggling to find the rent and food money out of their dwindling earnings. And while some families managed to live off the land for their food and fuel, they still needed cash for clothes, medicine and other essentials.

The better cottages and gardens were occupied by middle-income people such as the master of the local charity school, the carpenter and the tailor who relied on several sources of income. Henry Milstead supplied vegetables to the poorhouse while earning a salary as the vestry clerk and being paid to draw up the annual poor-rate accounts. George Bodle's friend, one of the unpaid parish overseers called Henry Mason, acted as both undertaker and carpenter, owing no doubt to his skill in making coffins, while William Blacknell sold flour and ran the local post office. David Rice combined running the Plume of Feathers with cultivating 1 acre and 2 roods of arable land.

Farther up the scale came the Budgen family of Manor House Farm next to the church, whose patriarch John, another one-time churchwarden and part-owner of a bank in Woolwich, had died in 1828. Unlike George Bodle, Budgen was classed as gentry. Then there was the magistrate, the Reverend Dr Samuel Watson. Like George Bodle, the wealthy Cleeve brothers didn't merit a place on *Pigot's Directory*'s list of local gentry although they occupied property and land worth several times that of Dr Watson. None of them, however, was worth as much as George Bodle.

Early on the morning of Wednesday, 6 November, Henry Mason received the call on his services he had been expecting. Old George Bodle was dead. Mason made his way over to the farmhouse, and there on the drive he found Middle John and Samuel Baxter huddled in conversation. Baxter told Mason that he had heard something very serious: Middle John was to be the next victim. When Mason asked Baxter what on earth he was talking about, Middle John said that Mary Higgins had overheard Young John saying there was but another that 'they' wanted out of the way, and that was his father. Mason thought the events of the past few days must have placed a strain on the young woman's imagination, but Baxter assured him that whatever Young John might or might not have said, he, Baxter, didn't for a minute think the boy capable of such a deed.

The three then went into the house. Whatever had gone on at Bodle's farm, something was clearly amiss, and Baxter decided that they had a duty to report the rumours. Young John had left Plumstead the previous Saturday evening within hours of the family falling ill and was staying with his sister, Mary Andrews, and her husband Thomas, at their coffee shop in Clerkenwell. Middle John's wife Catherine had written to her son and daughter as soon as George died summoning them back to Plumstead, but Baxter thought in the circumstances that it was better not to warn the young man, and he told Middle John to intercept the letter before it reached the mail coach.

The men then made their way into Woolwich to call on John Butler and find out what he thought about how George Bodle had died. The surgeon told them that he had already reported his concerns to the magistrate. What he did not tell them was that he had been collecting forensic samples from the day he was first called to the farm and that those samples had by then been handed over for analysis to Michael Faraday, professor of chemistry at the nearby Royal Military Academy.

When Middle John and his friends left the surgeon's house they walked to the home of the Reverend Dr Watson, where Middle John again recounted his story of how Young John had boasted about murdering his grandfather and was now plotting to kill him. Watson immediately sent for Mary Higgins, who, according to Middle John, had overheard everything. Even before Middle John and his friends had arrived, another disturbing piece of information had reached the magistrate's ears, and John Watts, a friend of Young John's and the schoolmaster at Kipling's charity school, was already on his way to be questioned.

Watts told the magistrate that two weeks earlier, at about seven in the evening, he had gone with Young John to Joseph Evans' chemist's shop in Powis Street, Woolwich, where his friend had bought two packets of arsenic. Who had tipped Watson off about this incident is a mystery, but by this time the magistrate had heard more than enough. He wrote a note to PC James Morris – 'a very active constable', according to the *Maidstone Gazette* – dispatching him to Clerkenwell to take Young John Bodle into custody.

Events in Plumstead now began to move fast. A terrified Mary Higgins was brought before Watson for questioning, while Constable Morris duly climbed aboard the Woolwich omnibus to Charing Cross, an arrest warrant in his tunic pocket. At the same time, as the law demanded, a body of seventeen 'good and lawful' local men convened as a coroner's jury in the large upstairs room at the Plume of Feathers.

John Ward, a prosperous farmer who had once given Middle John some work until the arrangement ended badly; Samuel Harwood, a relative of one of Young John's friends Jeremiah; Peter McDonald, the parish overseer, and fourteen other local men were told by the summoning officer that they had been called 'to appear here this day to enquire for our sovereign Lord the King when, how and by what means George Bodle came to his death'.

*

At twenty-four, Charles Carttar was only a year older than Young John Bodle but already installed as coroner for West Kent, a position he was to occupy for almost fifty years, his career off to a rather more promising start than that of the man about to have his collar felt in a Clerkenwell coffee shop. Carttar had taken over the post on the death of his father Joseph just eighteen months earlier.

The office of coroner, from corona or crown, dates back to medieval times. By the nineteenth century, most of the 300 coroners in England and Wales were lawyers who held the role for life, provided, in theory at least, they did nothing to disgrace themselves. Upon being informed of a suspicious death – often by a magistrate – the coroner had to summon between twelve and twenty-four men to be sworn in as jurors. The jury would first view the body and then hear from witnesses before deciding whether to send a suspect for trial at a crown court or to dismiss the allegations as unsubstantiated. The inquest was often held in tandem with the magistrate's investigations.

In 1751 the coroners' fees had been set by law at £1 for all inquests deemed to be 'duly held' plus travel expenses of 9d a mile, but this had led to arguments about what was meant by duly held. The Justices of the Peace, who held the purse strings, tended to the view that the coroner should become involved only when there was clear evidence of violence, but the coroners argued that it was their job to investigate every death that was unexpected and unexplained.

By the time Carttar junior stepped into his father's shoes, the system was badly flawed. First, coroners were largely unaccountable, with many remaining in office despite considerable blunders and, in several cases, obvious senility.

When Charles Lynn stood trial for the murder of Abraham Hogg at Aylesbury in 1825, such was the state of the paperwork that Mr Justice Gaselee received from the coroner that the judge declared, 'the gentleman must either be advanced in years or ... labouring under some mental infirmity'. As the gentleman in question was not in court, Gaselee said if the coroner himself did not feel that he was incompetent to hold his situation, he hoped 'some other gentleman present would recommend him to retire from the duties of his office'.

The future president of the Royal College of Physicians, John Ayrton Paris, worried that it was becoming increasingly difficult to find candidates with the necessary refinement and sophistication: 'As the office is attended with many unpleasant duties, gentlemen in these nicer times have shrunk from its performance and it has consequently fallen into disrepute; and too frequently into low and indigent hands,' he complained.

Over the years inquests had turned into something of a spectator sport, often with the body on public display while both witnesses and jurors fortified themselves with drink for the task ahead – this was not surprising, given the coroner's custom of using a room in a pub as a courthouse. 'The coroner frequents more public houses than any man alive,' said Charles Dickens. 'The smell of sawdust, beer, tobacco smoke and spirits is inseparable in his vocation from death in its most awful shapes.' The practice of convening in a pub stemmed from the idea that an inquest should take place somewhere well known and accessible, but upholding the dignity of the law in such surroundings often proved a struggle.

In his magazine *Household Words* Dickens went on to publish a more serious attack: 'If there appeared a paragraph in the newspapers, stating that her Majesty's representative, the Lord Chief Justice of the Queen's Bench, had held a solemn court in the parlour of the Elephant and Tooth-pick, the reader would rightly conceive that the crown and dignity of our Sovereign Lady had suffered some derogation. Yet an equal abasement daily takes place without exciting especial wonder. The subordinates of the Lord Chief Justice of the Queen's bench habitually preside at houses of public entertainment.' This distinction between the way different types of cases were tried was characteristic of Britain as a 'thoroughly commercial nation'. In a debt case when 'The Goddess of Justice' was called upon to decide whether Jones owed Smith five pounds, she was surrounded with 'the most imposing insignia', but when she was ruling on why 'a human spirit had been sent to its account', she was 'thrust into the Hole in the Wall, the Bag o' Nails or the parlour of the Two Spies'.

The article described an inquest in London's Drury Lane where the body was on view at the scene of the crime, a baker's shop, a few doors around the corner from the tavern where the inquiry was being held. At the shop, it appeared to be business as usual: 'There was the same tempting display of tarts, the same heaps of biscuits, the same supply of loaves, the same ranges of flour in paper bags as is to be observed in ordinary bakers' shops on ordinary occasions. Yet the mistress of this particular baker's shop lay dead only a few paces within, and its master was in gaol on suspicion of having murdered her.'

Much as *Household Words* railed against the lack of

dignity, so Thomas Wakley, the campaigning doctor, MP and founder editor of *The Lancet*, turned his guns on the coroners themselves, largely because they were pronouncing on medical matters without possessing any medical knowledge or being required to pay the slightest attention to anyone who did.

In 1823 John Ayrton Paris recommended that a medical assessor should be on hand to assist the coroner. Wakley, never one for half-measures, went farther. 'If medical men are best qualified to discharge the duties of coroners why not appoint them in place of those low incompetent persons?'

And the careless, unquestioning way that juries often pronounced cause of death came in for more criticism. 'If no troublesome person be present to disturb the even tenor of the way, and no unlucky juryman of sufficient intelligence to confuse the affair by putting an awkward question, the coroner tells the jury what they ought to do, the jury do as they are bid, and everybody gets away in good time for dinner or supper,' said John Gordon Smith.

Not all juries were so compliant, though. Occasionally one of Smith's 'jurymen of sufficient intelligence' made a stand. At an inquest in Bath in the 1840s, the jury insisted on having an expert analyse the victim's body along with the food he had last eaten and the medicine he was taking. They also tried to choose the surgeon for the post-mortem, which was by then their legal right. They got their way in the first matter and William Herapath, a leading toxicologist from Bristol, was duly tasked with carrying out the tests, but they were not allowed their choice in the second. So when it emerged that the reason the deceased's large intestine was not made available to Mr Herapath for analysis was probably because rats had been at the body, one of the jurors remarked: 'There must have been great neglect somewhere then … What would have prevented all this would have been if we had been allowed to choose our own medical attendant.'

In 1829 John Gordon Smith tried to stand for the coronership of Southwark and the City of London but was turned down because he lacked the traditional Oxbridge background. The following year the coronership for East Middlesex fell vacant, and this time Thomas Wakley put himself forward for election.

As it happened, at that time Wakley was involved in an inquiry into the death of a young Irish woman called Catherine Cashin, allegedly at the hands of a fashionable Harley Street quack called St John Long. Catherine's family asked Wakley to represent them at the inquest because with many wealthy, well-placed patients lining up to testify to Long's skills, they feared a cover-up.

Mrs Cashin and her daughters Catherine and Ellen had been lodging at the house of a couple called Roberts (or Roddis, as some accounts have it) in Mornington Place near the present site of Euston Station. Catherine accompanied Ellen to an appointment with Mr Long and while she was there the practitioner, who had no medical qualifications, said he could see just by looking at her that she also needed treatment. He then applied a poultice to Catherine's back. The resulting open wound between her shoulder blades became steadily more inflamed over the next few days while Catherine was racked with pain and nausea. After the young woman had passed a particularly dreadful night, Mrs Roberts left her in the care of a nurse while she slipped down to breakfast. 'A bell rang violently, I immediately ran upstairs and saw the deceased in the agony of death,' the landlady said. She then caused another sensation when she told the coroner that Ellen, who had been subjected to the same treatment, had died that very morning, just a few hours before the inquest opened.

A surgeon who examined Catherine's body found a large wound on her back that looked as though it had been produced by a scorching heat, 'for instance if a piece of red-hot iron, nearly the size of the crown of a hat, had been applied for about a quarter of an hour'. No substance was more likely to have done this than some strong arsenical preparation, he said.

In a perfect illustration of Dickens' point, the opening of the Cashin inquest had been interrupted by a drunken juryman, 'his conduct so boisterous as to distress the witness exceedingly and to put a stop for a time to the solemn investigation', according to one reporter. The coroner had seemed quite prepared to go on, however, and it was left to Catherine's brother-in-law to demand the man be removed as he was clearly in no fit state to carry out his duties. At this, the drunk became even more obstreperous and it took the beadles nearly fifteen minutes to overpower him and drag him away.

St John Long was sent for trial and appeared in court wrapped in a greatcoat, looking 'much agitated', and despite the roll call of the rich and titled who trooped into the witness box to pay tribute to his skills, he was found guilty of manslaughter. Mr Justice Park then fined him £250, which Long paid in cash on the spot, after which he was free to go home and carry on killing people if he so chose. And so he did. Just months later he stood trial at the Old Bailey once again, charged with the manslaughter of a Mrs Lloyd, who died in almost identical circumstances to Catherine Cashin. Extraordinarily, he was found not guilty.

The publicity surrounding the Cashin case and the outrage over the sentence did Wakley's bid for the coroner's post no harm – he was said to have been reluctant to get involved at first in case he was accused of using the tragedy to his advantage. Initially an outsider in the contest, in the end he was defeated by just 136 votes in a poll of 7,204. But in 1839, Wakley's time came: he was elected coroner for West Middlesex where, true to form, he proved a controversial, reforming and conscientious choice.

One of Wakley's most withering attacks had been on Charles Carttar's father. In 1829 Joseph Carttar had presided over the inquiry into the death of a three-year-old boy called William Adams. William was treated by one of the charitable Kent Dispensary's apothecaries for what was described vaguely as a complaint in the lungs. The child was given two grains of calomel and two of antimony every four hours, resulting in the boy ingesting 30 grains of each over the course of three days.

Calomel is a deadly compound that can destroy the gums and intestines before finally killing its victim from mercury poisoning. Antimony is a semi-metallic chemical element, again potentially calamitous. While both remedies did more harm than good, experienced doctors usually avoided killing their patients by prescribing much lower doses than those given to little William. In his case 'violent salivation ensued, the gums ulcerated and sloughed, the mouth and cheeks mortified [that is, the flesh became gangrenous]'.

The apothecary who ordered this toxic regimen did not see William again for four days and when he did, he declared that he could do no more. He had, it might be thought, done quite enough already. The mother then turned to the parish surgeon, who took one look at the lad and sent urgently for more expert colleagues, but while they were on their way, young William died.

At the subsequent inquest Carttar senior failed to press the surgeon about the likely cause of death and did not call the mother at all. The jury brought in the common verdict of 'death by the visitation of God', but they added a rider that there had been culpable neglect on the part of both the apothecary and also the mother, who, forced to scrape a living by labouring in the fields, had left the child either alone or in the care of neighbours. Carttar senior struck out the jury's comments and reprimanded them for exceeding their duty, adding that the apothecary's actions were no concern of theirs. Medical men who were acting for free out of the goodness of their hearts could not be expected to give much attention to non-paying patients, he said. He consequently changed the verdict to 'Died by the visitation of God and not in consequence of the neglect of any person or persons'.

The Kent Dispensary asked its medical committee, whose members included John Butler senior and Dr Sutton, to look into the matter. In view of the coverage in 'the public daily papers', the officials felt that 'such reports, if silently passed over by the governors, may eventually prove highly prejudicial to the interests of the charity'. The committee concluded: 'It appears that he [William Adams] was seen at least seven times in eight days, that Mr Brown [the apothecary] did prescribe the proper remedies for the complaint under which the child was labouring, watched the effects of his remedies and continued his visits so long as he entertained the slightest hopes of his recovery.' Even so, Brown resigned, saying he had accepted the office of medical assistant for Lewisham 'on the express condition that his personal attention to the patients was not constantly expected'.

Wakley was outraged. First he lashed out at charity hospitals, infirmaries and dispensaries, calling them human slaughterhouses, founded mostly by quacks where 'scenes of blood' were 'frequently exhibited within their walls', and then he turned his guns on Carttar senior, who, it turned out, had something of a conflict of interest in that he had held the paid position of secretary to the very dispensary in question for thirty years.

The coroner failed to say by whom 'it could not be expected that a medical man should bestow much of his time on pauper patients', Wakley remarked, but 'If these words were uttered by Mr Carttar, there is no doubt that he is a very fit person to fill the office of secretary to the Kent Dispensary and of nearly all the hospitals, infirmaries and dispensaries in the kingdom, but that he is certainly a very unfit person to hold the office of coroner and such a booby ought not to be continued in it another hour.' Such was the system, however, that the booby in question not only continued unchallenged in office for the remaining three years of his life but was then succeeded by his son. On 3 May 1832 Charles was elected unanimously to the dispensary post at a salary of £39 19s 1d a year. Two days later he was confirmed as the new coroner for Kent and was now about to preside over the inquest into the death of George Bodle.

9

A Very Active Constable

PC Morris arrived at the Andrews' coffee shop in St John Street, Clerkenwell, on the morning of Wednesday, 6 November 1833, a magistrate's warrant in one pocket and a pair of handcuffs in the other. The policeman wasn't expecting trouble; everyone knew John Bodle for a mild-mannered, easy-going sort. Still, it was hard to predict what a man fuelled by fear and panic might be roused to. So when Morris spotted his quarry hovering nervously in the back parlour and stepped forward to make an arrest, he was relieved to find that, far from making a fight of it, Young John collapsed on the floor in a faint.

Sir Robert Peel had set up the Metropolitan Police in 1829, and by the 1830s areas outside London had begun to follow suit. The Metropolitan Police Act of 1829 defined the Metropolitan Police district as an area within about a 7-mile radius of Charing Cross. This was extended ten years later to take in all of Middlesex and those parishes in Surrey, Hertfordshire, Essex and Kent within a 15-mile radius. (The City of London was an exception and still is.) Woolwich fell into this second stage but the new force couldn't arrive too soon for the inhabitants. In 1838 they held a public meeting to discuss petitioning the government for better policing. 'When it is considered that there are 20,000 inhabitants, six or eight night watchmen and only one police officer by day, some further protection is necessary,' said the *West Kent Guardian*.

At the time of George Bodle's murder, however, Woolwich – and therefore Constable Morris – was still operating under the old parish system of policing, which consisted of a rather ramshackle collection of often drunken and bungling, and occasionally senile, parish constables and watchmen.

In 1823, Charles Carttar's father Joseph had pronounced one of the watchmen on his patch unfit for office and had him sacked. Thomas Hawley's behaviour certainly fell short of what the parish expected – although it was not exactly unheard of. Carttar senior was holding an inquest at the Green Man, Blackheath, into the death of James Smith. Smith had been shot in the chest in front of his wife Jemima when he disturbed a burglar in the early hours of the morning. Hawley was clearly drunk as he gave his evidence, and when Carttar asked why he had waited ten minutes before investigating the pistol crack he admitted hearing, Hawley said, first, he wasn't allowed to deviate from his beat and, second, if he had responded immediately he too might have been shot. Had Hawley run the 130 yards to Smith's cottage as soon as he heard gunfire, he would probably have caught the murderer, Carttar told him. It then emerged that while Hawley and his fellow watchman were cowering by their boxes, waiting until the coast was clear, the Smiths' fourteen-year-old servant, Robert Papworth, had challenged the murderer and chased after him before setting off alone in the dark to fetch a surgeon.

Despite being described in the press as a 'very active constable', James Morris also turned out to have the weakness that was seemingly endemic in his job. As Young John was being revived in an upstairs room, Morris took the opportunity to down a glass or two of Mrs Andrews' rum. This marked the start of what was to become a farcical thirty-six-hour drink-fuelled odyssey. On leaving St John Street the policeman hired a cabriolet – a two-wheeled lightweight carriage – and the pair bounced their way to the Cross Keys coaching inn in Gracechurch Street near the Bank of England, where they each had a glass of ale chased down with brandy and water. At the Cross Keys Morris searched Young John and took some silver from his pockets, along with a key. The key, the young man told him, was to a trunk in his bedroom, and there the constable would find some packets of white arsenic. They then took an omnibus from the City to Woolwich and made their way to the Plume of Feathers, where Morris finally produced his man before Coroner Carttar.

While constable and prisoner were on their way, Carttar, the jury, two police officers and a gaggle of reporters had travelled the few yards along the main road to Bodle's farm to see the old man's body. Inspecting the corpse was always the jury's first duty. What remained of George Bodle was laid out on the bed awaiting scrutiny. His hugely swollen abdomen, 'proceeding from the internal inflammation', as the man from *The Times* explained to his readers, marked as clearly as a wound the seat of the trouble. While the visitors were in the house, they also heard a short account of the day the family fell ill from the maid Sophia Taylor, who was still recovering and too weak to travel to the Plume of Feathers.

Back at the inn, John Butler gave a brief outline of his part in the events and the coroner then tried to take evidence from the young cowherd, Henry Perks, about who had filled the kettle that Saturday morning. He quickly gave up in exasperation, saying he had doubts about the lad's sanity and ordering him out of the room.

Then a surprise witness was called. James Marsh, described as a 'practical chymist', came before the court to announce that he had tested the remains of the Bodles' coffee and found that it definitely contained arsenic. How much he couldn't say. Marsh's testimony took what the *Morning Post* called the feverish state of excitement to a new level and Coroner Carttar decided the court needed 'another scientific gentleman to examine the infusion and fluid, and the deceased's stomach'. That, of course, meant a post-mortem.

At that point in the proceedings Young John, who had been waiting for some time in the passage with PC Morris, was brought in. He seemed fully recovered from the shock of being arrested, leading *The Times* to pronounce him 'perfectly firm and unembarrassed'. 'You have been taken into custody on suspicion of causing the death of your grandfather but no evidence has yet been adduced to implicate you,' Carttar told him. 'I wish to inform you that, if you think proper, you may put any questions to the witnesses. You are not obliged to put any questions nor to say anything to criminate yourself. Recollect that what you say will be taken down and, if necessary, used in evidence against you.'

'I will answer any questions,' the young man replied.

'You may do as you think proper. Have you any professional adviser to appear for you?'

'No, sir.'

'Then I will endeavour to see you fairly done by,' the coroner promised.

The next witness was the Woolwich druggist Joseph Evans. Evans had known Young John for about twelve years, since school, in fact, he said. A person who was known to the druggist was more likely to be served with arsenic than a stranger, Evans told the court, and the prisoner knew that. Young John had been to his shop about a fortnight earlier with the schoolmaster, Watts, and then again on his own, either on Thursday, 31 October, or Friday, 1 November, the day before the family fell ill. Each time Young John had bought arsenic. On the first occasion when he was with Watts, 'a conversation arose relative to the destructive powers of arsenic and the prisoner mentioned that a number of his fowls and among them a guinea fowl had been destroyed by rats,' Evans said. 'I did not weigh it but gave him about half an ounce wrapped up in paper on which was the word "poison" in large letters.' On the second occasion he had purchased about the same quantity of poison. The Bodles' buildings and lands were certainly troubled by rats but, according to Henry Perks, who had managed to impart one or two pieces of coherent information before being waved away by the coroner, they were normally killed with traps and ferrets, not arsenic.

Throughout the first half of the nineteenth century, newspapers, medical publications, experts like Robert Christison and Alfred Swaine Taylor, and nervous members of the public were protesting about the number of times arsenic found its way into species other than rodents and insects, whether by accident or design. The government's chief statistician, William Farr, remarked that it was 'questionable' whether arsenic killed more rats than people. Certainly it was cheap and readily available – in the 1840s Alfred Swaine Taylor would complain that the going price for between a half and one ounce of white arsenic was still just 2d, while larger quantities could be had for a shilling a pound.

But regulation was a long time coming. A *Punch* cartoon from 1849 entitled 'Fatal Facility, or Poisons for the Asking' shows a little girl in a shop, so young that she is only just able to peer over the counter, saying: 'Please, mister, will you be so good as to fill this bottle with laudum [laudanum] and let mother have another pound and a half of arsenic for the rats.' The shop assistant, who is designated 'A duly qualified chemist' but appears to be in his teens, replies: 'Certainly ma'am. Is there any other article?'

There was without doubt a need for an efficient pest killer at the time: householders waged a constant battle against the rats, mice, fleas, lice, cockroaches and bedbugs that threatened to overrun their homes. Rats were a particular menace, ruining crops, emerging from the sewers in packs, scurrying around under floorboards gnawing at the wood, fouling food and drink and attacking people in their homes. In the eighteenth century the problem worsened as the traditional black rat – the so-called 'old English rat' associated with the Great Plague – that people had come at least to know if not love was ousted by a bigger, paler, more aggressive species, known either as the Norway or Hamburg rat, according to which region was being blamed.

So countless packets of deadly white powder were to be found lying around in sheds and kitchens across the country, often worryingly close to bags of sugar and flour. White arsenic was mixed with flour or oatmeal and placed next to mouse and rat holes; stirred into lard or soap and rubbed on to furniture and mattresses; added to water for washing down floors and left about the house in solution in saucers; while farmers used the stuff in industrial quantities to protect their crops and as a sheep dip. *The Complete Vermin Killer*, a household manual from the 1770s, blithely advises mixing arsenic and butter into a paste and then adding wheat or barley meal and honey, resulting in a killer confection guaranteed to attract children as readily as rodents. The author does add the warning: 'As this is a strong poison you must use it with caution and always wash your hands afterwards.'

And then there were the inevitable mistakes made over the counter when druggists – 'however ignorant, an individual unable even to sign his own name', complained the president of the Royal Pharmaceutical Society – were allowed to sell bacon, butter and cheese from one side of their shop and poison from the other. The Society also took issue with a proposed definition of a pharmacist as someone involved 'in the vending and compounding of drugs and medicines'. The trouble with that, the Society said, was that it encompassed all sorts of undesirables, including 'grocers, oilmen and hucksters ... patent medicine vendors, quacks, mountebanks, bone-setters, medical herbalists, farriers and cow doctors'.

In 1845 Peter Watkins, a druggist from Clerkenwell, stood trial for manslaughter after selling a cab driver called

William Watts an ounce of white powder for his upset stomach. Watts mixed the stuff in some warm water, gulped it down and immediately began to complain of a burning sensation in his throat and stomach. Then the vomiting began. Watkins had given the man tartaric acid instead of tartrate of soda. The two bottles labelled 'Acid. tart.' and 'Sodae. tart.' in tiny print stood next to each other on the shelf and Watkins' brother Charles testified that on the night before the accident, in a hurry to shut the shop and go home, he had put them back in the wrong order. The jury took pity on Watkins and acquitted him.

The government's first attempt to impose some sort of national control over the sale of dangerous substances had been in 1819, with the 'Bill for Establishing Regulations for the Sale of Poisonous Drugs and for the Better Preventing the Mischiefs arising from the Inattention or Neglect of Persons Vending the Same'. This proposed that 'No person or persons whatsoever shall vend or expose to sale, or offer, or keep ready for sale, any white arsenic, arsenic acid, corrosive sublimate [mercuric chloride, then used to treat syphilis], acetate of barytes, nitrate of barytes, oxalic acid, sugar of lead, Goulard's extract of lead, prussic acid [cyanide], tartar emetic, solid opium or laudanum, without a printed label and the word "poison" being affixed on and to every bottle, phial, packet, box or other inclosure containing such drugs and medicines, not only while in the possession of the vendor thereof, but that such vendor shall also affix the like label upon every bottle, phial, packet, box or other inclosure, at the time such drugs and medicines are delivered to the person or persons purchasing the same.' And arsenic and oxalic acid would have to be coloured, arsenic with carbon, before being sold. Infringements were punishable by a fine. Goulard's extract – also known as extract of Saturn or sugar of lead – was a lead-based lotion championed by the eighteenth-century French surgeon Thomas Goulard, who used it externally to treat inflammation in cases such as burns, sprains and rheumatism. In 1861 in her *Book of Household Management* Mrs Beeton was still recommending it for sore eyes, and she gave her own recipe for making it at home.

As legislation went, the 1819 curbs were hardly draconian, but the chemists and druggists objected that they would interfere with trade without achieving the desired effect. Instead they wanted a voluntary code of practice: no arsenic, oxalic acid or corrosive sublimate to be sold without a printed label stating what it was and with the word 'poison' clearly displayed. They also declared that no one should sell poison who was not 'of sufficient age and experience to judge of the importance of the great caution necessary in avoiding the sale of them to improper or ignorant persons'. Their argument won the day: the Bill was dropped and, despite continuing rows, recommendations and deaths, nothing more was done until a series of arsenic murders in the 1840s provoked a further panic, resulting in the 1851 Arsenic Act. All of this, of course, was of little comfort to George Bodle.

When Charles Carttar adjourned the Bodle inquest pending forensic reports, Constable Morris wanted to know what he should do with Young John. The coroner was loath to lock the young man up before the jury had come to a verdict but was equally reluctant to leave him free to run off back to Clerkenwell or anywhere else he might choose.

So when, with uncharacteristic concern for his son, Middle John asked whether Young John could remain in Morris's custody at the policeman's house, and he, Middle John, would pay for his keep, Carttar agreed.

Morris then took the prisoner to his home in Woolwich, stopping off on the way for a glass of gin and peppermint at the Mortar Inn and also allowing the young man to visit his grandmother at the Bodle farm. Young John went into Ann's bedroom and swore to the old woman, who through some miracle was recovering, that he had done her husband no harm.

The morning after he arrested Young John, Constable Morris went to the cottage and searched the young man's room, opening a trunk with the key he had taken from John's pocket at the Cross Keys. Inside was a bottle containing about an inch and a quarter of liquid, another bottle of what looked like some kind of ointment, and two – or possibly three (there was later some dispute about the matter) – packets of white powder, two of them marked with the words 'Poison' in large letters, one of them open. The policeman put the bottles and packets into the large pocket where he carried his handcuffs and set off for Evans' chemist's shop in Powis Street. On the way he ran into a few of his friends, however, was sidetracked and spent the rest of the day on a pub crawl of Woolwich and Plumstead, including seven hours in one bar with a local butcher called Richard Bullock.

Not surprisingly Constable Morris was later unclear about what he had been doing and where, but he did remember one port of call. 'I was obliged to go into the Red Lion to take shelter from the rain, where I stopped some time, as it continued to rain,' he explained. At some stage while cowering in the bar for fear of the elements and sinking several pots of ale, Morris bent down to tie up his bootlace and managed to drop and break one of the bottles. When he did finally venture out of the Red Lion, it was only to progress as far as the Mortar, where he switched from ale to gin and peppermint and entertained his companions by handing round the evidence. A seed dealer named William Osborne amused himself by rubbing some of the powder over his chin, then spilling it down his trousers.

Finally, at about eight o'clock that evening, PC Morris arrived at his intended destination, Joseph Evans' shop in Woolwich, and showed the chemist what remained of the exhibits after they had been passed around by a group of drunks and dropped on a bar-room floor. The policeman had managed to collect a few drops of the liquid from the broken bottle and presented this in a tumbler. Morris was fairly steady on his feet, although he 'appeared excited, was a little fresh and talked loud and thick', Evans said. 'But he was in a fit state to attend to business and appeared to know what he was about.'

Although the druggist was hardly in a position to identify the dribble of liquid in the bottom of the glass – it was later sent for analysis – he did agree that the packets looked very much like the ones he had sold to Young John. He knew them by the marks on the paper. One, miraculously, appeared to be in much the same condition as when he had sold it, but the other contained only 20 of the 218 grains Young John had bought. 'If it originally did contain as much arsenic as the former one, a considerable quantity is missing,' the chemist stated, although whether it had ended up on Mr Osborne's chin and trousers or in George Bodle's coffee was a matter for the jury.

The next morning at around half-past nine Evans was opening up his shop when he found what looked like a piece of screwed-up paper on the counter next to where Morris had been sitting. It turned out to be a small packet containing about fifteen grains of a white powder that looked very much like arsenic. Not wanting to be bothered with it, and for some extraordinary reason not connecting it with Constable Morris, even though no one had come into the shop to buy anything similar since the policeman's visit the night before, Evans tossed it on to the fire. 'I do not leave arsenic carelessly about,' he was to say. Not until a couple of weeks later, when he heard that Young John had admitted to having three bags of arsenic in his trunk, did Evans remember throwing away just such a packet and real-ise the only way it could have got there was if Morris had left it behind. At that time Morris himself wasn't entirely sure what he had removed from the young man's bedroom, although he was later to insist that he only ever had two bags.

If James Marsh was right when he told the coroner there was arsenic in the coffee, then this was clearly no accident – a crime had been committed and one so cold blooded and careless of human life as to indicate the work of a psycho-path. For either the plan had been to wipe out the entire Bodle household, and it was hard to see a rational motive for this, or the murderer had been prepared to see everyone perish in order to kill his intended victim.

That only George had died must have been chance; the

dose was possibly too low to kill the young and strong, and Ann Bodle had drunk only a few mouthfuls, as opposed to George's half-pint bowl. And if the killer knew about Mrs Lear's custom of taking the dregs to her daughter, and it would have been hard not to, then he or she must also have known that the entire Bing family in their little hovel, including two-year-old Mary and four-year-old Ann, could have died too.

If George was the intended victim, then the murderer had been lucky in achieving his goal. Deed done, though, he or she was then doubly unfortunate, first that a provincial surgeon like John Butler had been astute enough to recognise arsenic poisoning when he saw it, and secondly that the authorities had decided to pursue the case, for neither detection nor investigation could be taken for granted in 1833. As the magistrates were coming under increasing pressure to save money, the cost of pursuing such an inquiry often overrode the demands of justice and even the need to stop a killer on a spree.

So it was that in another seemingly peaceful little village, Happisburgh in Norfolk, a serial arsenic poisoner was able to pursue his hobby unchecked, largely because it was cheaper to disregard the growing death toll among his friends and family than to ask a few obvious questions. By the time Jonathan Balls was exposed, he had dispatched his wife, one of his children, at least eight grandchildren and possibly his father and mother, as well as several neighbours and the odd stranger who happened to be passing through. The killings stopped only when Balls took matters into his own hands and committed suicide, fittingly through a large dose of arsenic.

It then emerged that the local justices had been told to 'proceed conservatively' in calling for inquiries because of the expense. The forgoing of inquests for the sake of frugality was 'nothing less than ... a premium upon poisoning', the *British Medical Journal* commented.

It was unsurprising, then, that at the end of the first day of the Bodle inquest a row broke out between Coroner Carttar; William Nokes, the lawyer appearing on behalf of the parish; Charles Parker, the Greenwich solicitor who had drawn up George's new will; and two of George's executors, George Wassell and Henry Mason, about who should pay for the post-mortem, the further forensic tests that the coroner had ordered, and any criminal prosecution that might result. Parker didn't see why Bodle's executors should have to meet what was bound to be a considerable bill. He had understood that the parish was paying. Mr Nokes then jumped up to say that he could not recommend that the parish foot a bill of anything between £100 and £150, particularly when so many of the villagers were very poor.

Carttar thought Ann Bodle wouldn't object to paying, or at least he hoped not, seeing that this matter concerned the death of her husband, and he couldn't think that Middle John would do anything to obstruct an inquiry into the death of his father. He then asked Parker, 'as a gentleman seriously concerned in the investigation of the supposed murder of an individual, whether they should stand still for want of evidence'. Parker replied he wasn't concerned with the plight of Bodle's widow or his son; he was representing the executors, who, if ordered to pay, would have to do so out of their own pockets for there was no provision under the will for them to claim it back from George's estate.

At an impasse, Carttar sent a deputation of Nokes, Parker, Wassell and Mason to talk to Ann Bodle. Half an hour later, Nokes returned on his own. Mrs Bodle said she was not able to pay the expenses and point-blank refused to have anything to do with a prosecution should it be necessary to send anyone for trial. Nokes had also spoken to Bodle's son-in-law Samuel Baxter, who the previous week had accompanied the old man when he went to change his will. Baxter too refused to support any investigation but, like Ann Bodle, he declined to say why.

By this time, many people in the inquiry room were beginning to wonder whether anyone else in the Bodle family might have something to hide – even Carttar observed that the conduct of the family was 'exceedingly unnatural'. In the end, the coroner decided he had no choice but to ask the parish to pay and Nokes said they could probably recover the money by suing the family. With the matter settled for the time being, William Nokes now set about organising the medical and scientific investigations that the authorities hoped would send George Bodle's killer to the scaffold.

10

The Introduction of Irritating Matter

Charles Carttar changed his mind about needing one
other scientific gentleman on the case. After hearing
the testimony of the druggist Joseph Evans, he announced:
'It will be necessary to have the opinion of three medical
gentlemen who will be required to make a careful examination of the coffee, the fluid and the body of the deceased.'
The inquest could go no farther until the jury had heard
evidence on the cause of the deceased's death, he decided,
and at six o'clock the proceedings were adjourned.

For centuries doctors had been going into the witness box
to help the courts decide on matters such as how a blow
had been inflicted, whether a baby had been born dead or
was the victim of infanticide, or whether a woman had been
raped. They also appeared in civil cases, for example if a
husband or wife wanted a marriage annulled on grounds
of non-consummation, when the argument could turn on
establishing the impotence of one partner and/or the vir-
ginity of the other. But it was only towards the end of the
eighteenth century that these random contributions began
to develop into a specialist branch of medical science.

The true forerunners of medical expert witnesses in England were the panels of 'responsible matrons' or midwives who examined women under sentence of death who were 'pleading the belly'. From at least the fourteenth century, women who were 'quick with child' – that is, the foetus's movements could be felt – had their executions delayed until after the birth. In practice the sentence was often commuted to life imprisonment or transportation. 'I pleaded my belly, but I am no more with child than the judge that tried me,' says a woman in Daniel Defoe's *Moll Flanders*.

No distinction was made between juror and witness until the mid-seventeenth century, when the two roles began to separate. Before the seventeenth century, juries were chosen for their knowledge of the background to a case and of the character of the defendant. By definition, they were local men and could advise the visiting judge about the circumstances of the crime. By the end of the century the specialist jury had disappeared and witnesses were allowed to give evidence only on matters of fact, leaving a gap when the court needed specialist knowledge to guide it. The problem was solved by allowing those who were designated experts to give their opinions, unlike ordinary witnesses.

By the nineteenth century, though, the courts had adopted a less than deferential stance towards doctors and entering the witness box could often be something of an ordeal. In Continental Europe medical experts were officers of the court, paid set fees and their written reports had special status. Under English common law, doctors had no such privileges. They gave their evidence orally like everyone else, often under subpoena and unpaid, and there was no obligation for the jury to take notice of a single word they said.

A judge at the Old Bailey in 1786 had made this abundantly clear. Two carpenters, William Stone and John Neale, were accused of killing a stonemason called John Harrison in a drunken fight over money outside the Coach and Horses in the Strand. All three men were working on the Somerset House building site a few yards along the road. An apothecary called John Gadd saw Harrison four days after he had been knocked down several times and kicked. He found his patient delirious with a fever and suffering from a head wound. Gadd attended Harrison for the next three days until he died, but was determinedly equivocal in the witness box.

'What was the opinion you formed of that man's case?' he was asked.

'My opinion is very doubtful as to the cause of his death.'

'Are you of opinion that the fever and the delirium were occasioned by the wounds and bruises he received?'

'That is doubtful to me; I cannot say.'

'Have you a real doubt in your mind what that fever was occasioned by?'

'There is a probability it might be occasioned from intoxication, from passion, from anger; or it might be from the falls he received; it is very doubtful to me.'

At this point Mr Justice Gould lost patience. 'The court are to form their judgment from the evidence they have heard, and if they are of opinion that the death arose from the ill usage given by that prisoner at the bar, certainly the fact is proved ...' he told the jury. '... suppose a surgeon was to say, where it was manifestly clear that the injury was the occasion of the death, that he was of opinion it was not; is that to blind the jury? They are not to be hoodwinked or blinded; though not persons of professional skill, they are endued with common sense.' And when the jury asked to hear from the surgeon that Gadd had consulted for a second opinion, the judge stepped in again: 'Surgeons are called only to assist your judgment, they are not the people to determine this or any other case; you are to exercise your own judgment.' Even so, Stone and Neale were found not guilty.

As well as having his testimony ignored, the doctor could also expect to be tied in knots by fast-talking lawyers. 'Is there any object of dread, paramount in the eye of the medical practitioner, to the witness box?' asked England's first professor of medical jurisprudence, John Gordon Smith. Medical witnesses sometimes looked even more terrified than the accused, he thought. In court the doctor was on his own; he couldn't run home to his books or round to a colleague's house for advice. 'There he is and there he must remain,' Smith said, 'and [if he acquits himself badly] endure the scrutiny and displeasure of the bench, the brow-beating of the bar, the scorn, the laughter or contempt of the audience, the discontent of his friends and the exposure of the public press, with all the consequences that may follow to his reputation and fortune.'

But while some doctors found it difficult to give an absolute opinion, and restricted themselves only to what the medical evidence allowed them to say, this was not the line taken by all such witnesses. In fact some medical men – and judges and juries too for that matter – needed little encouragement in jumping to conclusions, particularly where unmarried women charged with infanticide were concerned. The celebrated surgeon and man-midwife William Hunter, brother of the equally famous surgeon and anatomist John, warned about 'the evidence and opinions given by physical people who are called in to settle the questions in science which judges and jurymen are supposed not to know'. In infanticide cases, he said: '... some of us are a little disposed to grasp at authority in a public examination by giving a quick and decided opinion where it should have been guarded with doubt'.

In particular Hunter highlighted the practice of putting the baby's lungs into water. If they floated, then they must have air in them, which proved – so the thinking went – that the child had drawn a breath and therefore had been born alive. But there were other quite innocent explanations, Hunter claimed. The mother might have blown air into her baby's lungs in an unsuccessful attempt to revive the child, for example. And if a baby made but one gasp before dying, the lungs would also swim as readily in water as if the child had been strangled at birth.

But while deploring the lawyers' brow-beating, many medical men had to admit that some of their colleagues deserved a drubbing. In 1795, George Hardwicke was accused of shooting a man with a blunderbuss. The surgeon George Burroughs examined the body and told the court that the victim had had a head wound.

'On your opinion, as a man of science in the profession, it was indicted by a gun shot?' he was asked.

'It was,' Burroughs replied.

'I don't know whether you washed the wound or saw it washed in your presence?'

'I did not. I directed it to be done; it was not done in my presence.'

'You did not extract anything from the wound?'

'I did not.'

'When you say it was your opinion, did you form that opinion from the appearance of the wound only or from such information as you may have received?

'Merely from the appearance of the wound.'

'You had, I suppose, heard at least, what passed?'

'Yes.'

'That did not assist your judgment?'

'It certainly had a very different appearance from what we call an incised wound or a contused wound.'

'You have seen many gun-shot wounds?

'No, I have not been in the habit of seeing many of them.'

At that point one of the Old Bailey regulars, the famous William Garrow, remarked: 'Thank God there are not many practical surgeons of that cast.'

The judge then intervened to ask what side of the head the wound was on. 'Upon my word, I have almost forgot on which side it was,' exclaimed Burroughs. 'I think it was the left, but I am not positive of that.' The accused was acquitted.

As well as the risk of public humiliation, doctors had another cause of resentment over being called as witnesses. Even if they managed to escape with ego and reputation intact, there was no guarantee of being properly paid for their time or even being paid at all.

The profession was mollified somewhat in 1836 by the passing of the Act to Provide for the Attendance and Remuneration of Medical Witnesses at Coroner's Inquests.

This piece of legislation owed much to the editor of *The Lancet*, Thomas Wakley, who was by then MP for Finsbury. It granted doctors a more official status at inquests, even if not as lofty as that of their Continental colleagues, but more importantly it guaranteed them a fee: one guinea for giving evidence and another guinea if a post-mortem was required.

By the early nineteenth century, the post-mortem had begun to play a central role in helping to establish cause of death. This was partly because of doctors' growing understanding of the processes, or pathology, of various diseases, and the changes that these diseases brought about in the body. Now as well as the outward symptoms, such as small-pox's dreaded little sacs of pus under the skin or the racking cough and bloody handkerchief that announced tubercu-losis, doctors found that the state of internal organs such as the liver, heart and lungs had important information to impart. And if the post-mortem helped explain death from natural causes, it could surely do the same where the grim reaper had been given a bit of a nudge.

And so, on the night of Thursday, 7 November, the day after Charles Carttar adjourned the inquest, three doctors gathered around the decomposing body of George Bodle in the farmhouse. Set out before them were the tools of the trade: knives, saws, scalpels, hooks and in this case a glass jar. In charge of the procedure was Samuel Solly, lecturer in anatomy and physiology at St Thomas's Hospital Medical School in London. Assisting him were the surgeon John Butler and a Woolwich physician called Francis Bossey.

Samuel Solly, the expert hired by the Plumstead parish lawyer William Nokes, was the twenty-eight-year-old son of a City of London merchant from a family of religious dissenters. Appointed to his post at St Thomas's Hospital earlier that year, in 1836 Solly would publish a major work on the human brain. He also coined the term Scrivener's Palsy, or writer's cramp, to describe a disabling condition that became widespread among civil servants in the 1830s; it was blamed on a new steel pen nib that had recently replaced the quill in government offices.

Working alongside Solly and Butler was twenty-four-year-old Francis Bossey, like Butler a member of an established Woolwich medical family. By authorising Nokes to hire both Solly and Bossey, Coroner Carttar was sparing no trouble or expense.

Watched closely by Butler and Bossey, Solly first studied the intact body, his eyes and fingers inching their way over the skin. He found no external signs of violence and the only abnormality was the swollen abdomen that the man from *The Times* had remarked upon. Satisfied then that any injuries were entirely internal, Solly began to cut.

The old man's innards had much to tell. The membrane lining the gullet was inflamed and the inflammation also ran down the windpipe and into the lungs. The lungs themselves were congealed and contained more than the normal quantity of blood. 'On cutting into them a large amount of fluid poured out, which is unnatural,' Solly wrote.

When he opened the stomach he found the mucous membrane lining was unusually thick, and on the left side it was dark brownish black, while the right side of the stomach was 'a pale, dirty pink with yellowish spots in many places'. There were also 'small depressed points, about the size of the extremity of the probe, round which points were slightly elevated rims, not flattened'. The first and middle sections of the small intestine were also inflamed. Arsenic scours the surface of the stomach and intestines, causing inflammation and then ulcers. After a few days, the arsenic reacts with the hydrogen sulphide given off by the decaying body, resulting in patches of yellow orpiment (arsenic trisulphide).

The doctors carefully collected the contents of the old man's stomach and intestines and put them to one side. If they were not absolutely scrupulous in how they handled and stored them, then a clever defence counsel might argue that any traces of poison were the result of contamination. 'This may be regarded as a very remote presumption but nevertheless it is upon technical objections of this kind that acquittals follow, in spite of the strongest presumption of guilt,' Swaine Taylor warned. He knew of just such a case where a doctor had thrown stomach contents into a jar borrowed from a nearby grocer's shop where poisons were sold.

Solly then moved on to George Bodle's head, and here he found an unusual amount of serum under the dura mater (the outer of the three membranes covering the brain and spinal cord) – about an ounce in the brain's ventricles, the four communicating cavities contiguous with the central canal of the spinal cord. He then turned to the major organs. The old man's liver was granulated and the pericardium, or membrane around the heart, was sticking slightly to the surface. These conditions were long-standing, however, and they were not what had killed George Bodle.

In Solly's opinion, none of these findings alone could account for the old man's death, but the anatomist thought that some 'irritating matter' had definitely been taken into the stomach. (The different appearance of the two sides of the stomach could have been because the ingested matter had coated one side more than the other, perhaps because the old man had been lying on his side in bed.) The doctor thought the farmer's partial loss of sight and his inability to stand could have been caused by inflammation of the brain, which supported Butler's theory of poisoning, but, Solly said, only in the most severe cases was it possible to tell from the appearance of the body that poison had been taken. Nor could a post-mortem alone prove cause of death. In fact, the symptoms were often decisive, Solly said, and in this case, neither food nor an internal complaint could have been responsible for the state of George Bodle's stomach.

Solly's conclusion, then, was that death was caused by the 'general disturbance of the constitution produced by the introduction of some irritating matter into the stomach'. 'I think it probable that the same cause would not produce death in a young person,' he said. 'The inflammation would be greater in a young person but the disturbance to the constitution would be less than in an old one.' And given that the inflammation extended from the back of the mouth and throat right down the gullet and into the stomach, as well as down the windpipe to the lungs, he believed that the irritating matter was arsenic.

Butler agreed with Solly, although as a provincial surgeon he would have needed more than his fair share of confidence to argue with an expert from one of the country's leading medical establishments. The physician Dr Bossey was a little more cautious. In his opinion, George Bodle had certainly suffered from an irritant taken into the stomach but he would only say that this was probably, not certainly, the cause of death. Bossey also thought that only two substances, tartar emetic – potassium antimony tartrate, then used in medicine as an emetic – or arsenic, could have affected Bodle's stomach in this way.

With the post-mortem complete, John Butler took the old man's stomach contents in a large clean jar, sealed it firmly and, in accordance with Swaine Taylor's guidance, delivered it personally to the chemist James Marsh at the Woolwich Arsenal.

On 8 November, the day after the post-mortem on George Bodle, the *Morning Post* published a story not carried by any of the other newspapers. The person suspected of the 'fiendlike deed' was the victim's own grandson, 'whose character in the neighbourhood stands in very bad repute', the paper claimed. Young John had arrived at the farm 'at an early hour and long before the servant had arisen' and proffered his services to light the deceased's fire, boil the kettle and clean the hearth. 'He was desired to go away but he would take no denial,' the *Post* claimed. After alleging that Young John had handled the coffee pot that contained the poison, the paper said he had then 'absconded and directed his course towards London'. Thither Constable Morris pursued him 'and captured the fugitive near Smithfield'. The accused had made several purchases of arsenic 'from divers chemists in the vicinity'.

What could have inspired such a piece, so full of errors, all highly prejudicial to Young John? One possible clue lay in the last paragraph: 'His father, in speaking of the lamentable occurrence, has been heard to declare his opinion that it was the intention of the prisoner to make him the next victim.' While waiting for the inquest to resume, Middle John appeared to be telling his version of events to a wider audience.

11

I Never Saw Two Things in
Nature More Alike

When Coroner Carttar required an expert to analyse John Butler's samples for the presence of arsenic, Michael Faraday, professor of chemistry at the Royal Military Academy a mile down the main road in Woolwich, was the obvious man to consult. Here after all, on the doorstep, was one of the most renowned chemists in the country.

Faraday had taken the part-time post teaching young Royal Artillery and Royal Engineer officers at the academy in 1829. By 1833, he was also Fullerian professor of chemistry and laboratory director at the Royal Institution and therefore somewhat over-qualified for the Royal Military Academy job. But the man whose pioneering work on electromagnetism was to make him one of the most celebrated scientists in history had spent the previous three years doing highly boring research into optical glass at the insistence of his boss Humphry Davy, the inventor of the miner's safety lamp. Faraday accepted the Military Academy chair in order to gain some measure of financial independence so that he would never again be forced to accept such tedious projects as the glass work.

By the time of the Bodle case Faraday had made several court appearances as an expert witness, mainly in civil cases giving evidence over matters such as patents and pollution. But he turned down the Bodle investigation. The professor was preoccupied with matters that, from a scientific point of view at least, were of rather greater significance than hunting for arsenic in vomit and coffee. Just four days before the inquest, Faraday had written to a friend: 'I have ... been so deeply engaged in experimental investigations of electricity that I have not read a journal, English or foreign, for months. My matter in fact overflows, the doors that open before me are immeasurable. I cannot tell to what great things they may lead and I have worked neglecting everything else for the purpose.'

And for someone of Faraday's eminence this small job was hardly worthwhile, carrying only a small fee compared to the sums he was used to charging for his evidence in civil cases. The professor was not unhelpful, though, and he handed over the job to the local man who acted as his assistant for his chemistry lectures, James Marsh.

A portrait of Marsh painted a few years after the Bodle case, when he was probably in his late forties, shows a pleasant-looking man with a high forehead, finely shaped nose, marked chin and brown eyes. Small swathes of thick dark hair are brushed forward on to his cheeks at the side of his temples in the Napoleonic style. Dressed appropriately for the sitting in a high collar and thick black stock, Marsh wears the serious expression befitting a decorated man of the Royal Society of Arts, although his ruddy, strongly made face is at odds with the popular image of the scientist in the lab.

Marsh's main job was as surgeryman and dispenser of medicines at the Royal Arsenal, in other words he was an assistant to the resident doctor there, but he was also an exceptionally gifted scientist and engineer. The previous year his employers, the Ordnance Board – the government department that ran the Woolwich Arsenal – had given him a £30 bonus for an invention that was of great benefit to the Royal Navy. On a warship, the interval between taking aim and firing a cannon was critical to hitting the target because the ship's motion made it impossible to keep the object lined up for more than a split second. Improvements to the firing method had speeded up the process but there were problems: the force of the explosion often caused the copper percussion tube that contained the firing powder to shoot out, wounding the gunner, and the gun itself often misfired. In tests on board HMS *Excellent* at Portsmouth, James Marsh's crow's quill percussion tube proved not only safe to use, but out of 900 rounds fired not a single one missed its target.

James Marsh's upbringing and his early life remain something of a mystery. Marsh was just too poor, too respectable and not famous enough to have left much trace. Until the late 1830s information about people of his class was largely restricted to a one-line entry in a parish register of baptisms, marriages and deaths, and even those scant details were sometimes omitted or wrongly recorded. Some sources give his date of birth as 1789 and describe him as an Irish physician who studied in Dublin and practised medicine there for the early part of his life; others claim he studied in Dublin but that his subject was chemistry rather than medicine. There is no record of this, however, and it is hard to believe that a physician – a doctor at the top of the nineteenth-century medical hierarchy – or a formally trained chemist would end up working as an assistant to the resident doctor. One possible explanation for the suggested Irish connection is that James has been confused with Henry Marsh, who was indeed a Dublin physician, born in the 1790s. The truth is almost certainly that James Marsh was born in Kent in 1794, as the census states, probably in or near Woolwich in humble circumstances, and that he went to work at the Royal Laboratory at Woolwich Arsenal when he was about twelve years old.

The Arsenal was originally a massive weapons factory dating back at least to the 1500s when Henry VIII ordered ordnance stores to be built on the banks of the Thames. The following century, factories making brass cannon were set up alongside the warehouses, and towards the end of the 1600s the Royal Laboratory producing explosives, fuses and shot was moved from the Tilt Yard at Greenwich to Woolwich. The Brass Foundry, the Carriage Department, the barracks of the newly formed Regiment of Royal Artillery and the Royal Military Academy for the training of 300 artillery and engineer officer cadets all followed. And close by the Arsenal, stretching for a mile along the river front, were the Royal Dockyards, again established by Henry VIII in the sixteenth century.

By 1833, then, Woolwich had a centuries-long history of service to the business of war. It was a garrison of over four thousand men and a training centre for cadets along-side a massive cacophonous machine for the churning out of weapons, and attached to it all was a lively town where attorneys, doctors, land agents, undertakers, corn merchants, boot-makers, butchers, grocers, straw hat manufacturers, blacksmiths and all the other dealers in the stuff of civilised life offered their goods and services. The eighteenth-century parish church of St Mary Magdalene stood to the west; a rabbit-infested common to the south; small shops and cottages lined the grubby lanes and more elegant buildings edged the main square. The academy was an impressive piece of architecture and the 1,200-foot-long barracks was described by one contemporary directory as 'one of the grandest public edifices in the Kingdom'. Woolwich, though, remained defined by the great blocks of warehouses, factories and shipyards along the river, and from time to time the Ordnance Board paid compensation to local farmers for sheep or cows that strayed into the path of the firing range.

In the heyday of the Peninsular and Napoleonic Wars, the Arsenal and the docks employed nearly ten thousand skilled workers and thousands more unskilled men and women. A a respite from foreign wars in the 1830s, however, meant the production of ships and weapons at Woolwich had fallen away: peace, at least as far as the local economy was concerned, was bad for the town. In 1811 over six hundred boys were working as labourers in the Royal Laboratory, receiving between 6d and 1s 3d a day depending on their size and strength, sweating to complete an order for 10 million ball cartridges, although when they reached the halfway mark, the Ordnance Board sacked 400 of them. James Marsh was probably one of those 600. In 1846 his wife Mary was to claim that he had spent forty years in the Arsenal's service, while the Royal Laboratory director,

James Cockburn, described him as having been 'brought up in this department'.

At some stage in his adult life Marsh met James Achindachy, a soldier in the Royal Horse Artillery, and the two became firm friends. In 1833 Achindachy was still in the army but he was later to become the landlord of the Queen's Arms in Artillery Place, Woolwich. Achindachy was to be at Marsh's bedside when he died and he is mentioned warmly in Marsh's will.

James Marsh was destined for greater things. In 1823, while he was still a labourer, he received a national award, the Society of Arts' (later Royal Society of Arts) Large Silver Medal along with 40 guineas, for his work on electromagnetism. When the Society of Arts was established in 1754, it had launched a series of competitions for inventions, discoveries and artistic endeavour with prizes in the form of medals and money. The scheme had flourished from the start and soon committees were set up to preside over six categories: agriculture, manufacturing, chemistry, mechanics, colonies and trade, and the polite (fine) arts.

Marsh's achievement in the chemistry section came about through his work in the early 1820s assisting another Royal Military Academy professor, the mathematician Peter Barlow – not with lectures but with experiments. Like Faraday, Barlow was investigating magnetism and electromagnetism and the invention that won Marsh the medal came about when Barlow asked Marsh for help in solving a practical problem that was hampering progress.

The relationship between electricity and magnetism was one of the hot topics at the time, exciting the attention of leading mathematicians, chemists and physicists both in

Britain and on the Continent. 'Exceedingly interesting and important' experiments on the subject had 'marked the principle [sic] track of philosophical investigation during the last four years', the Society of Arts said. The cumbersome and expensive equipment the work required, however, meant that the field was restricted to those few researchers fortunate enough to have access to the apparatus and made it nearly impossible to replicate the experiments in different parts of the world. In particular, researchers wanted to see whether variations in the earth's magnetism across the globe would produce corresponding variations in electromagnetism. The Society of Arts thought that Professor Barlow's mathematical laws of electromagnetism, 'however probable and consistent', should be tested by scientists on board His Majesty's ships around the world, so Barlow directed his clever assistant to put his mind to devising a cheap, portable apparatus to replace the complicated contraption then in use.

The Society of Arts met at its grand London headquarters in the Adelphi building off the Strand on 4 April 1823 to select that session's award-winners. The candidates included Mr J. W. Jeston of Henley-on-Thames, surgeon to the 36th Regiment of Foot, who was being considered for his improved method of collecting the juice of the opium poppy; a Woolwich man, Captain Dansey of the Royal Artillery, who had invented a type of kite to help stranded ships communicate with the shore; Joseph Amesbury, with a medical device to help a broken leg heal straight; Elisha Pechey of Bury St Edmunds, who had devised a new mangle to make washday slightly less laborious; and James Marsh.

Some of the hopefuls had sent in papers describing their

work but Marsh was there in person to demonstrate his exhibit, and he had an influential champion that day: Michael Faraday was on the committee. After a warm recommendation from Faraday, the committee concluded that Marsh's invention was 'capable of exhibiting all the known facts of electromagnetism and of enabling the professor [Barlow] to prosecute further researches in this interesting and important branch of natural philosophy', adding that the portability and reasonable price of Marsh's apparatus put it well within reach of most researchers and 'should peculiarly adapt it to the use of the traveller by land or sea'. It was indeed compact and practical, fitting into a box just 14.5 by 15 by 10 inches.

So Marsh, along with Mr Jeston and his poppy juice, received the Large Silver Medal, while in the mechanics section Captain Dansey's kite did even better, winning the Gold.

There was one discovery that did not feature on the list of awards for 1823, however. For while candidates were free to submit any work they chose, the Society also had some pet projects, and it had been urging researchers to tackle one in particular since 1821. This matter, concerning as it did the catching of murderers and the clearing of the wrongly accused, seemed more pressing than the easy extraction of poppy juice or an improved mangle. Yet it was also an especially tricky problem to solve, and it would be many years before anyone stepped forward to claim this prize.

Marsh's extraordinary aptitude and voracious mind were clearly noticed at the Royal Laboratory, for on 28 February 1824 the Ordnance Board accepted the recommendation of the director-general of the medical department, Sir John Webb, that James Marsh be appointed to the surgeryman and dispenser post, replacing James Plant, who had died two weeks earlier. During his later years many people, including his wife Mary and the director of the Royal Laboratory, James Cockburn, would describe Marsh as the Arsenal chemist. In fact, the Ordnance Board had abolished the post of ordnance chemist and assayist of metals in 1829 when the then incumbent was pensioned off. But although Marsh was denied the official title – and therefore a place on the staff establishment and the commensurate salary and pension – he certainly carried out duties far above his pay grade. As a surgeryman Marsh was paid just 4s 1d a day for a six-day week, or £63 18s a year. His role in assisting Faraday, which came about in 1830, earned Marsh a further 5s a lecture, or another £18 15s a year (compared with the professor's £200).

Soon after Faraday's appointment, the professor and Colonel Percy Drummond, Lieutenant Governor of the Royal Military Academy, were corresponding on the subject of the new chemistry assistant's pay. To help Drummond arrive at a fee, Faraday explained how he saw Marsh's role. 'I think experimental lectures owe all their value to the experiments and visual illustrations which are given in conjunction with the theoretical details and it will be my object to make these demonstrations as distinct and impressive as possible,' he wrote. 'I should think I should use Mr Marsh's time in first preparing them and then cleaning up and keeping order for about one and a half or two ordinary days for each lecture. I anticipate there would always be a little running work from the intention which I have of rendering the chemical establishment more and more complete as to its apparatus and preparation. Hence you will be able to judge what a man so occupied ought to have.'

The extra money must have been welcome, for James Marsh had married Mary Watkins, a farmer's daughter, in 1815, and by 1830 they had two daughters, Lucretia Victoria, known as Victoria, and Lavinia Berthia. A son, James Frederick, had been born in 1825 but with no record of him apart from an entry in the St Mary Magdalene baptism register, the presumption is that he died young, certainly before the 1841 census was taken. The colonel's final decision about what a man so occupied ought to have brought Marsh's total annual pay up to about £82 13s. It was still little enough on which to support a wife and two children.

In approaching Michael Faraday, Carttar and Nokes were being particularly diligent, perhaps because of the local standing of the Bodle family and the huge public interest that the case was attracting. In 1833 it was still common practice to call either on a local apothecary or a surgeon like John Butler to carry out any tests required, although there were clear dangers in relying on those with little or no expertise in the subject. In 1831, an anonymous author claimed that scarcely one in fifty doctors had the ability to determine 'whether a pudding or a mess of pottage be contaminated with arsenic or not'. 'Even the whole medical staff of our great metropolitan hospitals are often incapable of clubbing together a sufficient degree of chemical knowledge' for the task, he continued.

In 1821, a forty-five-year-old woman called Ann Barber was convicted of the murder of her husband after the apothecary-surgeon who carried out a post-mortem and tested James Barber's stomach contents declared himself satisfied that the cause of death was arsenic poisoning. John Hindle was even confident enough to estimate the amount of poison involved: 'I suppose he must have taken more than a dram,' he told York Crown Court. But after discussing his findings at some length in what sounded like a knowledge-able manner, Hindle announced blithely: 'I really cannot tell how many mineral poisons there are. I never applied the test before and never saw any other person apply it ... This poison is a subject I am very little acquainted with.' No one in the courtroom, least of all the jury who sent Ann Barber to the gallows, seemed to regard this as a drawback.

By the 1830s, though, universities and medical schools were offering new courses and soon an emerging group of specialists, the forensic toxicologists, were making their name as expert witnesses. Henry Letheby in London and William Herapath in Bristol in particular began to crop up regularly in criminal trials, along with Swaine Taylor and Robert Christison. Swaine Taylor became a familiar figure at the Old Bailey, but he too was famously caught out by the smoke and mirrors that seemed to bedevil the tests for arsenic.

Such was the nascent state of forensic toxicology in 1833, however, that James Marsh had never before carried out the arsenic tests. But while not formally trained, he was a talented chemist, and in any case his lack of experience with arsenic was academic as far as his first appearance before Coroner Carttar was concerned. There was no time to carry out the standard experiments so instead Marsh turned to an old, subjective test: he took some of the suspect coffee grounds, threw them on to the fire and sniffed the air to see whether or not he could detect a whiff of garlic. *The Times* described this as Marsh having 'analysed' the coffee, which was something of an exaggeration in the circumstances, and Marsh himself later explained that he had come to the conclusion that arsenic was present because of a 'peculiar smell', not through any chemical test. This was why, on the day after George's death, Wednesday, the 'practical chymist' was unable to give the coroner's court any idea of the quantity of poison present.

In fact, just a year earlier the Edinburgh toxicologist Robert Christison had recommended abandoning the garlic test altogether. In the third quarter of the eighteenth century it had been virtually the only test in use, he said, and one 'respectable author' (Johannes Daniel Reisseissen in 1781) had even recommended burning the victim's entire body if this was the only way to obtain the necessary evidence. The 'alliaceous odour' was not infallible proof, however, that arsenic was present, Christison declared, just as its absence proved nothing either.

Now, with the inquest adjourned specifically to allow time for the forensic work, Marsh was able to run the standard laboratory experiments, not only on all of Butler's samples, which by then included George's stomach contents from the post-mortem, but also on the packets of white powder, the drizzle of liquid from the broken bottle and the jar of ointment that Constable Morris had found in Young John's bedroom trunk and which, by some miracle, had survived the policeman's pub crawl. But although the tests that Marsh proposed were certainly more scientific than heating up the substance to detect a hint of garlic, they were still far from foolproof.

'Why do you believe it to be white arsenic?' the defence at the trial of Mary Blandy asked the prosecution's expert witness, Dr Anthony Addington – 'it' being the mysterious powder that the thirty-one-year-old heiress Mary admitted putting into her father Francis's gruel. The year was 1752, and the story, involving as it did a wealthy family, had fascinated the nation just as the Bodle case was doing now.

The lawyer Francis Blandy was a wealthy widower living in Henley-on-Thames in Oxfordshire, and among those vying for the hand of his only daughter was the Honourable Captain William Henry Cranstoun. If one contemporary view is to be believed, the captain was hardly the catch of the season: 'His stature is low, his face is freckled and pitted with the smallpox, his eyes small and weak, his eyebrows sandy, and his shape no ways genteel,' the description ran. 'His legs are clumsy, and he has nothing in the least elegant in his manner.' Even so, it was Cranstoun whom Mary chose and, as the son of a Scottish nobleman, he was initially acceptable to Francis.

The first sign of trouble came when it emerged that there was something of a drawback in terms of Cranstoun as a future husband: he was, in fact, already married, and there was a child. When a bailiff then turned up on the Blandy doorstep demanding to know the captain's whereabouts, Francis's opposition to Cranstoun became absolute. Not long after he made his views clear, the lawyer was attacked by steadily worsening stomach problems and after a few days he died. Mary then admitted that she had dosed her father's food with a white powder. She said Cranstoun had given it to her, saying it was a magic potion that would reconcile Francis to the marriage.

When Dr Addington, who was not a chemist but the Blandy family physician, was asked why he believed the 'magic potion' was white arsenic, he replied: 'This powder has a milky whiteness; so has white arsenic. This is gritty and almost insipid; so is white arsenic. Part of it swims on the surface of cold water like a pale sulphurous film, but the greater part sinks to the bottom and remains there undissolved; the same is true of white arsenic.'

Addington then described the first of the tests he had performed on the powder, the same test that James Marsh used on the coffee grouts. He threw the suspect substance on to red-hot iron. Arsenic 'rises entirely in thick white fumes, which has the stench of garlic and covers cold iron held just over them with white flowers', the doctor explained. In fact, it is the element arsenic, not arsenious trioxide or white arsenic, which produces the distinctive garlic smell when heated, but white arsenic can be reduced to its elemental state when it reaches a certain temperature, so Addington and Marsh may well have both smelt what they said they did. The 'white flowers' refers to the white arsenic that forms a deposit on the iron when the grey elemental arsenic combines with oxygen and cools.

Still satisfied then that he was dealing with arsenic, Addington moved on to what were known as the colour tests. He boiled 10 grains of the powder in 4 ounces of water, filtered it and poured it into five glasses. 'Into one glass I poured a few drops of spirit of sal ammoniac [sal volatile, or smelling salts]; into another some of the lixivium of tartar [a solution of potassium carbonate]; into the third some strong spirit of vitriol [sulphuric acid]; into the fourth some spirit of salt [hydrochloric acid] and into the last some syrup of violets [used to detect acids and alkalines].'

The spirit of sal ammoniac 'threw down' a few particles of pale sediment, the lixivium of tartar 'gave a white cloud which hung a little above the middle of the glass', while the spirit of vitriol 'hardened into glittering crystals' and the syrup of violets produced 'a beautiful pale green tincture'.

The doctor then reran the tests on some arsenic and found 'an exact similitude between the experiments made on the two decoctions'. 'They corresponded so nicely in each trial that I declare I never saw any two things in nature more alike than the decoction made with the powder found in Mr Blandy's gruel and that made with white arsenic,' Addington said. Mary Blandy was hanged.

By the time of the Fenning case in 1815, science had moved on a little. Twenty-year-old Eliza Fenning was accused of trying to kill her employers – Robert Turner, a Chancery Lane law stationer, and his wife Charlotte – and their household. Mrs Turner said Eliza had asked to be allowed to make some dumplings for dinner, saying she was 'a capital hand' at the dish. The dumplings were duly served along with rump steak and potatoes. Almost immediately the diners collapsed with burning stomach pains and vomiting. At half-past eight the local surgeon who attended them, Mr Ogilvy, decided that he needed a second opinion, and John Marshall was summoned from his home in Half Moon Street, Piccadilly.

The first sight that confronted Marshall was Eliza Fenning slumped on the stairs, apparently in agony. Pausing only to tell her to drink some milk and water, Marshall hurried up to see the family. Their bowels seemed to be 'in knots', and the surgeon tried to provide a little relief by massaging their stomachs with a warm flannel soaked in laudanum. Robert Turner appeared to be nearly 'in articulo mortis' (at the point of death), his swollen face exhibiting the true 'facies hippocratica', the look of someone about to die, as described by the Greek physician Hippocrates. Gradually, though, over the next few days, with the doctors prescribing large amounts of fluids with the aim of flushing the poison out of their systems, the household began to recover.

As was usual at the time, the prosecution centred mainly on motive, opportunity and the general behaviour of the accused. Charlotte Turner said Eliza had seemed resentful ever since she told her off for entering the room of one of her husband's apprentices half dressed, while much was made of Eliza's failure to come to her employers' aid when they were struck down. Her own speedy recovery was put down to her having eaten only a tiny portion of dumpling in order to avoid suspicion.

Called upon to be an expert witness, John Marshall described how he had washed out the dish used to mix the dumplings with a kettle of warm water. He'd let the liquid stand, then decanted it off and dried the sediment that he found in the bottom of the bowl. He was left with half a teaspoon of white powder. 'I decidedly found it to be white arsenic,' he announced. Charlotte claimed that no one but Eliza had been in the kitchen while the dumplings were being made, so the prosecution argued that if they did indeed contain arsenic then Eliza must have put it there. Robert Turner told the court that he had noticed the family's knives and forks were badly tarnished that night and he produced two of them for the jury to see. Marshall was asked whether arsenic would turn iron black. 'I have no doubt of it,' the surgeon replied. In fact, he was wrong.

The surgeon said that while he had based his initial diagnosis of arsenic poisoning on his patients' clinical symptoms, which was reasonable enough, he had then set about looking for scientific proof. The dumplings were clearly the most likely cause of the trouble, so Marshall had cut the remains into thin slices which revealed white particles thickly distributed throughout, he said. He then put a portion on to a polished halfpenny and held it on the blade of a knife over a candle. As it burned, it gave out 'most unequivocally the garlic smell'. When cooled, the upper surface of the coin was 'of a silvery whiteness occasioned by the fumes of the arsenic', in other words Dr Addington's white flowers.

Marshall next put a few grains of the white powder from the mixing dish between two copper plates, bound together with a wire, and poked them through the bars of a grate. On removing them red hot from the fire, the garlic smell was again 'highly perceptible', while, as they cooled, the plates gave off the white fumes typical of arsenic and again exhibited a silvery whiteness.

For the colour tests, Marshall, unlike Addington, had turned to an expert. Joseph Hume, a chemist based in Long Acre, Covent Garden, was credited with developing what had become known as the silver test some twenty years previously. Hume poured a solution of Marshall's powder into a glass and added silver nitrate. The 'silver test' produced a yellow cloud that gradually turned the transparent solution an opaque yellow, Marshall said, which 'beautiful and highly satisfactory experiment infallibly proved the powder to be white arsenic'. The chemist next introduced copper sulphate to more of the solution. This time the result was the grass-green precipitate copper arsenite known as Scheele's green.

Marshall had turned up at the Old Bailey armed with his samples, ready to demonstrate the experiments to the jury, but the court would not spare the time. 'I very much regret the want of an opportunity of proving … at least two of these experiments – the silver nitrate and copper sulphate – for these most eminently removed every doubt,' the surgeon said.

Concerns about the guilty verdict delayed the execution for three months but the Home Office finally declared the conviction safe, and Eliza died. When she leant over on the scaffold to whisper into the chaplain's ear, many people thought that terror for her soul had finally prompted a con-fession, but the Newgate 'ordinary' later revealed that, on the contrary, she was frantically confiding her innocence once again.

One of Eliza's most energetic supporters was a writer called John Watkins, who published a detailed rebuttal of the prosecution's case. Watkins calculated that if the lick of dough left in the mixing dish had indeed produced half a teaspoonful of arsenic then the four and a half dumplings that had been eaten must have contained a total of 1,800 grains of the poison. As 5 grains were enough to destroy any human being, the quarter of a dumpling that Mrs Turner ate would have killed ten people and the one and a half that her husband ate, 120 people. But since no one died, the only way of accounting for the large amount of arsenic Marshall had recovered was if the powder had been sprinkled over the dough after it had been made, rather than added during the mixing. Therefore Eliza was not the only person with the opportunity to commit the crime and Mrs Turner's insistence that no one had entered the kitchen while Eliza was preparing the mixture was irrelevant.

Watkins then posed a set of questions for Marshall: What became of the arsenic? Why wasn't it produced in court? How did the surgeon know it was arsenic? What became of the remainder of the dumplings – were they tested by feeding them to a dog or cat? Was the mixing dish the only vessel Marshall examined? Why didn't he examine the pot used to boil the dumplings? What became of the water they were boiled in? Did he ask where the water used to mix the dumplings came from? Did he inspect the vessel it was fetched in? Did he examine the milk can that hung in the kitchen? And the salt vessel from which the salt was taken? Did he examine the sauce?

On Marshall's performance over the knives Watkins had a wider point to make: 'It is not true that arsenic will produce the effects of blackness upon a knife. Was Mr Marshall aware that he was giving evidence and opinions upon oath, which from him as a professional man the jury would assume to be true?'

For John Gordon Smith, fighting to make forensics compulsory for medical students, the Fenning case was a gift. In one of his first lectures as professor, he produced two boxes, one containing a knife that had been left for ten hours in arsenic and the other a knife that had been left with pickled walnuts. In a dramatic 'open sesame', he then brandished an untarnished knife from the arsenic box and a blackened one from the pickles.

By the time James Marsh set to work on the Bodle case, testing for the presence of arsenic had been largely standardised to the reduction test and the precipitates. Reduction relies on the fact that when white arsenic is heated it loses oxygen, or is reduced. If this is done in an open-ended tube, the oxygen escapes, leaving metallic arsenic on the glass in the form of what looks like a mirror, a dark deposit with a dull sheen. If the 'mirror' is then heated gently, the arsenic is reoxidised and white arsenious oxide appears in the tube once more.

The precipitate tests worked in the same way as Addington's colour tests, but by 1833 chemists relied largely on three chemicals for their results: Hume's silver nitrate test, which produces a yellow precipitate when arsenic is present; copper sulphate, which Hume also used in the Fenning case and which produces Scheele's green; and hydrogen sulphide gas, which when bubbled through the solution produces yellow arsenious sulphide in the presence of arsenic.

It is unclear whether James Marsh used the precipitate tests on the Bodle samples – newspaper editors thought, perhaps incorrectly as it turned out, that readers were not interested in the technical details. If he did so, then it was only to corroborate the results of his main method: the reduction test, reducing the white arsenic to its elemental state and then reversing the process.

George Bodle proved something of a disappointment from a chemical point of view: Marsh found nothing remotely suspicious in the old man's stomach contents nor in his vomit (in any case, Mrs Lear later admitted that with so many people being sick, she could not be absolutely certain that the sample did come from George). Equally, the fur from the kettle that the cowherd Henry Perks had emptied into the garden yielded no results, despite Marsh examining it 'with great care'. After Henry's repeated scrubbing and rinsing, however, even if the kettle had originally contained a whole ounce of arsenic, it would probably have been impossible to detect it, Marsh said.

The crust in the coffee jar in the parlour cupboard that John Butler had so carefully preserved again produced nothing, and while the chemist thought he could detect a small amount of poison in the ointment in Young John's bedroom trunk – ointment which the accused had already admitted contained arsenic – he could not say for certain. The coffee from the Bings' cottage – which had given off the 'peculiar smell' when Marsh threw a sample on to the fire the night George Bodle died – turned out to be more rewarding, however, as Marsh was to explain.

Marsh's methods were slightly more refined than Addington's and Marshall's, but they were still rudimentary and imprecise; the difference was that Marsh realised that. In 1824, Robert Christison had urged caution regarding tests on organic matter. In most medico-legal cases, the analyst was working with pieces of the stomach and its contents, and the tests risked being 'enveloped in much difficulty and uncertainty'.

So for all Coroner Carttar's insistence on spending the parish funds on chemical tests, the fact was that even eighty years after Francis Blandy died, the best way of demonstrating death by arsenic poisoning was still by studying the victim's symptoms while he was still alive, backed up by the appearance of his digestive system and other organs at post-mortem.

So, the following Monday, 11 November, with the

doctors' autopsy and James Marsh's tests complete, the inquest into the death of George Bodle could resume. Early on, an excited crowd gathered outside the Plume of Feathers. The pub's upstairs room set aside for the proceedings was large, but no sooner had the coroner taken his seat than the court was packed, with people spilling out on to the stairs, straining to hear what was going on. The county magistrates, the Reverend Dr Watson, who had issued the warrant for Young John's arrest, and Mr Stace, attended, as well as 'several other influential residents of Plumstead, Woolwich, Greenwich and its vicinity', according to the *Morning Chronicle*. If they were expecting a thrilling show, they would not be disappointed.

12

She Would Not Risk
Her Soul into Danger

The Bodle inquest proved a marathon, running the entire week and late into the Friday night, with the newspapers recording the proceedings almost minute by minute. The likely murder of one member of a wealthy, respected family and the possible hanging of another had caught the imagination of the nation and the inexperienced young coroner was under pressure to get things right. Consequently Carttar had allowed the lawyers to question and cross-question the witnesses in the sort of detail more usual in a criminal trial, frequently intervening himself when he spotted an inconsistency. (He was not always to take such pains. Ten years later a newspaper would write admiringly of his ability to rattle through five inquests in an hour.)

The first witness was Middle John's servant, Mary Higgins. Mary had been kicking her heels in the poorhouse for the past week, by order of the magistrates, to stop her colluding with the other witnesses, in particular her master Middle John. William Nokes, the lawyer for the parish of Plumstead, now took her through her story. She began by describing how she had found Young John up and dressed, waiting by the fire that Saturday morning, and how he had set off in the direction of George Bodle's farm as soon as it grew light. Then she recounted what had happened that evening when Middle John came back from collecting his wages at the Bodles' farm. 'He said all the family were very ill and the doctor said they were poisoned. My mistress said "La, who would do such a thing?".' Young John, who had been in his room most of the afternoon, heard the commotion and opened his door, demanding to know what was the matter. She had told him, 'Master says Mr Bodle and all the family are ill and they supposed they are poisoned.' 'He made little answer and shut the door again,' she told the court.

Then came the evidence that had led to Young John's arrest. On the day the family fell ill, between dinner (as the country people called their midday meal) and tea, Mary heard the prisoner say he would not mind poisoning anyone he did not like. 'That was addressed to my mistress. My mistress said she would not risk her own soul into danger for anyone. The prisoner replied "Oh I would not mind. Only give me the stuff and you will see". My mistress said it was light talking.'

And there was more. 'I heard the prisoner say one day the previous week that he wished his grandfather dead, when he should have a thousand or a hundred a year – I don't know which.' With the average wage for maids such as Mary much lower than the £6 a year her London counterparts could expect, her vagueness about how much Young John stood to gain was unsurprising; the sums were beyond her comprehension. 'My mistress said "Lord, John, how can you talk so?" and the prisoner replied he should like his grandfather to die one day and his father the next. I said one should die one week and the other, the other. The prisoner replied "Ah, that will do".' Later Mary was to claim that she also heard Mrs Bodle wish her husband dead during that conversation.

And the day after the family fell ill, the sharp-eared Mary had apparently also overheard another rather mysterious snatch of conversation between mother and son. 'On Sunday night about nine or ten o'clock, after my mistress had returned from the deceased's house, she said to the prisoner: "I dare say the Baxters will have something to say of your going so soon in the morning for the milk". He replied: "Oh I dare say they won't. You will be the first to invent that". My mistress replied: "God knows John if you have anything about you, it is unknown to me".'

Nokes then turned to the day after the old man died. 'On Wednesday morning I got up very early by my master's desire,' the girl said. 'It is very unusual for him to desire me to get up so soon. At about five o'clock my master came into the kitchen. He said Sophia and Perks were against John for filling the kettle. I told him I heard Young John say he wished him and his grandfather dead and that he would not mind poisoning anyone he did not like.'

By now Young John had acquired his own lawyer, James Colquhoun, William Nokes' partner. The pair shared a practice in the elegant Rectory Place, Woolwich, where three years later the surgeon John Butler was to take a house. Between them the lawyers handled much of the legal work in Plumstead and Woolwich.

After her gentle handling by a sympathetic Nokes, Mary

was feeling much more confident than she was when first ushered into the room and was therefore quite unprepared for Mr Colquhoun. He started softly enough, asking her a few questions about the day her master had roused her at five o'clock and they had sat together in the cold kitchen in the candlelight with Middle John's wife and sons asleep upstairs. But then the lawyer's tone became increasingly aggressive, until finally he turned on her, suggesting that her entire testimony about Young John discussing poison was a pack of lies cooked up by his father. Mary promptly burst into tears and, according to the reporter from the *Maidstone Gazette*, 'cried bitterly and for a long time nothing could be obtained in addition to her former evidence'.

Mr Colquhoun then moved on to Mary's relationship with Middle John and his son. Wasn't it true that she was on terms of great familiarity with her master? No more than she ought to be. Wasn't it true that she had made advances to the prisoner and hadn't he complained about her behaviour? Not that she had heard. Didn't she keep knocking on his bedroom door? Only to wake him up in the morning. Didn't she keep going into his room when he was still in bed? Only to make the bed. Didn't she make him an apple-pie bed? (Placing the sheets in such a position as to prevent anyone getting into bed, the *Maidstone Gazette* helpfully informed its readers.) Only once, and she meant nothing by it. Didn't she go into his bedroom when he was undressed and didn't he tell her to get out? He wasn't undressed, he just had his jacket off, and she had left the room as soon as she had put the quilt on the bed. Mr Colquhoun then asked her to repeat a conversation that he said had taken place recently in Charlton Lane. 'It's a pity you've nothing better to do than put such questions to me,' Mary told the lawyer before bursting into tears once more. Having suggested Mary would do anything that Middle John asked her to do, including helping to destroy his son, because she was having a sexual relationship with one and had been rejected by the other, the lawyer finally let the gibbering girl go.

With revelations like these on offer, the room at the Plume of Feathers stayed packed 'almost to suffocation', according to one reporter, and an irritated Carttar repeatedly had to order the beadle to clear the doorway as the crowds were pressing so heavily against the table that the jury were nearly knocked out of their seats.

After the rows about who was to pay for Charles Carttar's team of scientific gentlemen, there was of course huge interest in what they would have to say. Dr Solly was the first to give evidence and he repeated his opinion that the farmer's death was due to a 'general disturbance of the constitution produced by the introduction of some irritating matter into the stomach', and that the irritating matter was arsenic.

Then James Marsh was recalled. By now he had the results of his laboratory tests and he told the court that he had found nothing conclusive in any of the samples – with one important exception. He was able to say that the 8 ounces of coffee that George Bodle had consumed would have contained quite enough arsenic to kill him. In the 5 fluid ounces of liquid coffee that Mrs Lear had fetched from her daughter's cottage and handed over to John Butler, Marsh had found between 4 and 6 grains of poison. The Bodles' kettle held about 7 quarts, or 265 fluid ounces, so Marsh had added arsenic to 7 quarts of water in the proportion of half a grain of arsenic to 1 ounce of water. The consistency of the mixture looked very similar to that of the Bodles' coffee, he thought, but on this point, he warned, he was 'obliged to speak with great latitude'.

Next into the witness box was Middle John, and from the outset the father of the accused continued where Mary Higgins had left off, with a narrative that seemed to be steadily lowering a noose over his son's head. Young John had come in late that Saturday morning, he told the coroner, and it was unusual for the lad to go for the milk; one of the farm boys usually fetched it. He had certainly never seen his son with a milk can. 'He was in and out of doors in the course of the day but where he went to I don't know.' He hadn't known about Young John's trip to London that day but then he never asked his son about his whereabouts. 'Sometimes he is out for two or three days and I don't know where he goes to. I don't trouble myself about it,' Middle John said.

As for his own movements over the crucial time, on the Friday afternoon before the family fell ill he had been working in his father's market garden adjoining his cottage, some distance from the farmhouse. He finished work at about five o'clock and went home. He had not gone to his father's house that night, nor had he walked past the gate. On the Saturday morning he had got up later than usual, at about 7.30, and gone straight to the common to see to the sheep, and had stayed there until about nine.

The coroner then turned to the day after George's death: was it true that he had told Mary Higgins to get up early that morning and that they had talked in the kitchen by candlelight? It was not, Middle John said. He was not in the habit of conversing with Mary Higgins 'more than what a master is in the practice of doing to a servant'. He rather thought he had been the first to get up that morning but he did remember seeing Mary at some stage. He then began to relate an exchange he said the pair had had, despite his earlier denial of talking to her. He had asked who fetched the milk the previous Saturday and she had told him Young John. He had had no particular reason for asking, he said: 'I cannot think what induced me. I never asked the question before.' When the coroner reacted to this with incredulity, Middle John changed his story. He had asked, he said, 'in consequence of something my father said on his death-bed'. He had also mentioned another matter to Mary, again for no apparent reason. He had asked whether Catherine and Young John had 'had any words' in the week and Mary said yes. 'I asked what it was about and she said it was about simmering something up in a cup before the fire. I did not ask her what was in the cup, as it did not concern me. She said my wife and John had had a good many words about boiling this stuff.'

Middle John then related a conversation that he had had with his dying father as he sat at his bedside holding his hand, which he claimed had prompted him to ask Mary Higgins about the milk. 'I asked him if he had any suspicion of who had done anything to him. He said he was satisfied it wasn't me who did the deed. He said "It was your son John and I am well convinced he did it".' As to why he had asked the question, Middle John said: 'Because it was reported about that he was poisoned and there was an oration about the streets.' 'Did you make any answer to it?' 'No I did not.' 'Why did not you reply to your father when he accused your son?' 'He was in a dying state, and being agitated I did not think of it any more.'

Mrs Lear had been at the other side of the bed when George Bodle accused Young John of murdering him, Middle John agreed, but he didn't think she could have heard the conversation. 'It was in a low tone of voice and I could only just hear what my father said. I was obliged to put my ear close to his mouth.' He had told Henry Mason, George Wassell and Samuel Baxter what his father had said before going to the magistrate, he said. But when the char-women, Mrs Lear and Mrs Wooding, were called they flatly contradicted his version of events. They were constantly by the old man's bedside as he died and if the conversation with Middle John had taken place, they would have heard it, the women insisted. Mrs Wooding added that at no time did Middle John lower his head to speak to his father. Baxter, Wassell and Mason were also to deny that Middle John had repeated this deathbed conversation to them.

At that a juryman, the Plumstead vestry overseer Peter McDonald, interrupted to say that he had chanced to meet Middle John the morning after the old man died and Middle John had told him they suspected that the farmer had been poisoned by one of his own family. Middle John looked across at the coroner: 'Mr McDonald asked me how my father was. I said he was dead and that it was reported about that he was poisoned and that I had strong informa-tion who did it.' 'Did you not say that you were in posses-sion of a plot and how it was to be transacted?' McDonald asked. 'He asked me if I knew who did it,' came the answer. 'I didn't mention a word about a plot.' But then turning to McDonald, Middle John added: 'I told you I suspected who it was, or that we were tracing it.'

George's trusted son-in-law Samuel Baxter then told the court that the day before George died he had told the old farmer that people were saying he had been poisoned but his father-in-law certainly hadn't confided any suspicions to him. Baxter added that on the Wednesday he had overheard Middle John scolding Judith Lear for saying that the family had been poisoned, which surprised him because by then it was common gossip.

In his capacity as an executor of George's will, Baxter was then asked about its contents. He told Coroner Carttar that as things stood Middle John was to 'receive property' on the death of old Ann Bodle. The old woman was very frail and suffering from many ailments, according to John Butler, so this event was unlikely to be long in coming. In fact Ann was to die three years later. Baxter went on to tell the coroner about the inheritance due to Young John, his brother George and their sister Mary Andrews. What Baxter failed to mention was that he and his eldest son William also stood to gain acres of farmland, while his wife Mary-Ann and their other children would inherit valuable stock.

On the strength of the contradictory stories about who said what and when, the coroner re-called Middle John and pushed him hard. On the one hand Middle John repeated the claim about his father's suspicions regarding Young John. But then he said that he had not repeated this to Baxter, Wassell and Mason and had told them only what Mary Higgins had said.

'You said in your evidence and I have it in my notes that you did make that communication to Mr Baxter, Mr Wassell and Mr Mason ... You said that Mr Baxter told you on Monday night that your father was poisoned and that it was reported about that he was poisoned. How can you make these statements agree?' said the coroner.

'It was in consequence of what Mr Baxter said that I asked my father if he thought he was poisoned and he said yes, he thought he was.'

'Why, there is another contradiction to your evidence. Do you know the situation in which you are placed?'

'Yes sir, I do.'

'Then what did you tell Mr Baxter?'

'Oh I hardly recollect now what it was.'

'The circumstance seems to have made very little impression on your mind. What did you state to him?'

'I told him what the girl had told me about my son talking about getting rid of me.'

Then it was the turn of Young John's lawyer Mr Colquhoun, who wanted to know more about Middle John's movements on the crucial Friday night before the family fell ill. He had indeed gone home at five o'clock as he had already testified, Middle John said, but when Mr Colquhoun pressed him, he admitted that he had gone out again, at about half-past eight or nine o'clock, either to the Plume of Feathers or the Green Man, where he probably stayed about two hours, but he could not remember exactly. And as for not going near his father's house, while he rather thought he had gone to the Prince of Orange that night, which was in the opposite direction to the farmhouse from his cottage, he could not swear to it. 'I do not recollect passing the gate of my father's house but if I did go to the Green Man I must have done,' he told the lawyer.

Carttar intervened at this point to ask whether he had got

up in the middle of the night and gone downstairs. 'I will not swear I did not,' Middle John said, 'but if such was the case it must have been to let my other son in.' He refused to be drawn any further but insisted that even if he had got up in the middle of the night, he had not left the house. 'The witness was interrogated for some time upon this point,' reported the *Morning Chronicle*.

On the second day of the reopened inquest, Tuesday, 12 November, the coroner and jury again made the short journey along the main road to Bodle's farm, where this time they inspected the kitchen and the parlour. The main purpose of the visit, though, was to hear from the widow. They were led into Ann Bodle's private sitting room and received with some formality by the old lady, sitting in an armchair by the fire and dressed in deep mourning. 'She appeared to labour under great infirmity occasioned by old age, but in full possession of her faculties and perfectly aware of the object of the jury's visit,' said *The Times*.

Despite that firm hold on her faculties – or, could it be, because of it? – Ann was to be of little use. She went through the events of the Saturday the family fell ill. She had seen Middle John briefly that evening when he came for his wages, and his wife that night when she came to see how they were, but she hadn't seen her grandson that day. Her husband did say he thought there was something in the water. He didn't mention his son or his grandson to her but then he was a man of little conversation at the best of times. She cried and 'appeared greatly affected' when she described getting off her sickbed in order to see her husband just before he died and how they had prayed together.

She described how the family had used up the last of the

coffee and the sugar that morning and how she had had to have lump sugar. She was fairly sure that her husband had not given his key to the coffee cupboard to anyone, nor could anyone have taken hers because she always put her pockets under her pillow when she went to bed. (By 'pocket' Ann meant a small cloth bag that was usually carried only by women who could not afford a leather purse. George and Ann probably regarded leather as an unnecessary extravagance.)

'You must let me have that key,' Carttar told her.

'But I cannot do without it.'

'But you have another.'

'No I have not. Mr Baxter has that.'

Samuel Baxter then produced a set of keys, one of which he admitted did indeed fit the coffee cupboard. He said he had taken it from his father-in-law's pocket only after the old man's death. Why, he did not say.

Ann was then asked about her husband's relationship with his grandson. The two were on fairly good terms, she said, and she had never known the boy to offend his grandfather. When asked why George had banned his grandson from working on the farm, she said it was simply that her husband had thought him not strong enough for the work. There had been no falling out; in fact after the incident the old man had paid the boy to help with the haymaking. She knew nothing about the dying George telling Mrs Lear that Young John was not to come for the milk any more and felt sure that if Mrs Lear had been ordered to turn the boy off the premises the charwoman would have mentioned it. In fact, the boy seldom came to the house and when he did he only ever went to the wash-house and then neither she nor her husband saw him because he always 'brushed off'. Mr Nokes pounced on this: 'What do you mean by saying that John brushed off and was not seen by his grandfather?' But the old lady had another innocent explanation: 'Because his grandfather thought he had no business there, hindering the girls.'

When the action moved back to the Plume, the onlookers were in for a further treat in the form of showdowns between Mary Higgins and her employers. If Middle John and Mary had been trying to collude, they failed wretchedly. Mary and Middle John's wife Catherine were brought in and made to face each other while Carttar read out Mary's statement that Young John had told his mother he wouldn't mind poisoning anyone who offended him. A shouting match broke out across the room between the two women.

'It is false as I am on my oath,' declared Mrs Bodle. 'The conversation never took place.'

'It did.'

'Speak the truth.'

'I do speak the truth. It did take place. I positively say it did take place.'

'And I as positively deny it.'

Carttar then came to the allegation that Young John had wished his father and grandfather dead. 'I swear such a conversation never took place,' said Mrs Bodle, 'and if it did it is more than Mary Higgins dare do to make any observation. I never heard my son say anything disrespectful about his grandfather in my life.'

Next was Mary's statement that when Young John returned home on the Sunday evening, she had heard Catherine say the Baxters would have something to say about his filling the kettle, adding, 'but, God knows, if you have anything, I know nothing about it', and John replying, 'You are more likely to do it.' 'All I said was, I advised him not to go for the milk, for the world was wicked and might make something of it,' Catherine said. And as for Mary Higgins' tale about mother and son quarrelling over the cooking up of a mystery potion over the fire, John had taken marrow from some beef bones and was melting it to make pomatum for his hair. Catherine said she had told him to stop, she would heat it up in the oven the next day, and he had told her not to interfere. Carttar asked whether, when Middle John came home on the Saturday evening, he had gone into the kitchen and said to the two women: 'There's a pretty to-do at father's: they are all poisoned.' 'He did say so,' Catherine replied. Carttar then turned to Mary Higgins and for once the women were able to agree: 'He did say that, or something like it,' Mary told the court. Clearly by then loyalty to Middle John was in short supply.

The tension mounted again when the coroner recalled Middle John and asked him whether, when the family up at the farmhouse still thought they were suffering from English cholera and before the surgeon John Butler had made his diagnosis, he, Middle John, had told his wife they were poisoned.

'False,' replied Middle John.

'What say you to it, Mary Higgins?' the coroner asked.

'He did say so, sir.'

'False, I did not,' shouted Middle John.

Carttar next quoted Mary's claim that on the night the old man died Middle John told her that she was to get up early the next morning when he called her.

'I am on my oath and I never said such a thing.'

'She said you called her at five o'clock on Wednesday morning?'

'It is the biggest lie that was ever said; I was not up that morning till between six and seven o'clock.'

'You sat down by candlelight to breakfast,' Mary interrupted, adding, 'How can you tell such lies?'

The coroner then asked whether Middle John had told Mary he was going to Woolwich that Wednesday morning, that he wanted her to go before the magistrate and that Sophia Taylor and Henry Perks suspected Young John. 'Lies all of it, on my oath,' said Middle John, at which Mary Higgins shouted again 'He did say so' and her employer replied, 'Why, you false wench, how dare you?'

Young John's lawyer, James Colquhoun, brought the confrontation to an abrupt and intriguing end by asking Middle John whether he'd ever been in Maidstone prison. The witness was clearly not expecting this. 'I don't know that I have any right [duty] to inform you,' was his first reaction, before deciding: 'I have no objection to telling you; it was for cutting hop bines and I was acquitted.' Ten years earlier, on 14 June 1823 at Kent Summer Assizes, John Bodle was charged that he 'with force and arms at the parish aforesaid unlawfully, maliciously and feloniously did cut certain hop bines … belonging to Moses Gratwick, to the great damage of the said Moses Gratwick, against the form of the statute, such case made and provided and against the peace of our said Lord the King, his Crown and Dignity'. Despite the evidence of three witnesses – Moses Gratwick, James Bellingham and William Hodges – Middle John had been found not guilty of this vandalism, but had been held in custody while awaiting trial. There are no details about how Middle John came to be accused but a clue might lie with the prosecution witness William Hodges, as later events would show.

Mr Colquhoun had some other matters to mention about Middle John's earlier dealings with the jury foreman John Ward.

'During the time you were in Mr Ward's employ, were you not charged with fraud for selling a horse?'

'He will have his money some day.'

'Wasn't there also something about a bill which Mr Ward settled for you?'

This time the witness decided that Colquhoun had indulged in enough character blackening for one day: 'I shan't answer you. I have no right to answer that question,' he told the lawyer.

~ 13 ~

Oh My Poor Mother

With tensions at fever pitch, at last Coroner Carttar called the star turn: the charming, lazy Young John Bodle, dressed in the latest fashions. Mr Colquhoun began by asking him to explain those early-morning visits he had made to the farm, which had begun so suspiciously just two weeks before his grandfather died. The story that emerged was one of innocent flirtations between three young people that had started with Young John telling his cousin, the cowherd Henry Perks, to ask Sophia for some cream when he fetched the milk and Sophia sending back a message that he could have some if he came himself. So began a pattern of helping the lively Sophia and the pretty, deaf-and-dumb Betsy with their chores; opening the shutters, filling the kettle, banking up the fire, churning butter and skimming milk in the cellar, along with 'gambolling', as he called it, 'capering around' in his grandfather's heavy old boots and his grandmother's bonnet and dress to make Betsy laugh, and waking Sophia up by banging on her window with a stick.

On the crucial day, Saturday, 2 November, he had got

up too early, mistaking moonlight for daylight, and had set off to fetch the milk, but then he heard a clock strike five and so he turned back home to wait for daybreak. Mary Higgins then came down to the kitchen and soon after he went to his grandfather's house. Here he met Betsy by the back gate. The kettle was standing outside the kitchen door and the wash-house door was open. He found Sophia in the kitchen cleaning the stove and asked whether he could do anything to help. She had said, 'You can do your old job if you like, filling the kettle.' He took the kettle to the pump in the yard, emptied out the water, refilled it and put it back by the kitchen door. He would have hung it over the fire but Sophia hadn't yet laid it.

He had been sitting down for a few minutes when some-one had knocked at the door, he said. It was a man beg-ging. He told the man he had nothing to give him because the family was not up. Sophia had seen the man and asked who he was, he said. (Sophia was later to tell the coroner she knew nothing about a beggar.) He was about to leave when Sophia complained that he hadn't finished his task so he had brought the kettle in and hung it on the crane. He then collected his milk and headed home. As he opened the shutters, he had heard the ringing of the Arsenal bell. At that one of the Woolwich magistrates, William Stace, inter-rupted to say that the Arsenal bell had not been rung on a Saturday since 1 October. Young John said that he was sure it was the Arsenal bell.

Asked whether he could account for the arsenic that Constable Morris had found in his trunk, he said yes, he used it to treat 'the itch'. He put half an ounce in a bottle, mixed it with 6 ounces of water and applied it to his skin rash with a piece of rag. He used that liquid during the day and at night an ointment, made up of 4 ounces of lard and half an ounce of arsenic, which he rubbed on with his fingers. He had been using arsenic in this way off and on for four years. 'Sometimes I did not use it for a month but when the disease was bad I applied it two or three times a day.' Previously he had used hellebore root (also poisonous), 'but I found more relief from the arsenic than anything else'. He had bought the poison from Mr Evans for that purpose. 'One packet I opened but the other I did not. I formerly kept a coffee shop in town but left London two years ago and brought some arsenic with me. The quantity I brought from London served me two years with other things I used.'

'The itch' was scabies, seen then, as now, as shameful because of its unsightliness and its association with dirty living conditions. The lumpy red rash is an allergic reaction to a mite, *Sarcoptes scabiei*, that burrows under the skin and lays eggs. These hatch out, setting up a cycle that is hard to break. The itchiness of the wounds provokes a mad scratching that often leads to a bacterial infection, making the skin even hotter and angrier. In the nineteenth century, there was no effective remedy, and the condition was recurring.

Young John's care over his skin was characteristic of a man who wore fashionable clothes in the depths of rural Kent and heated bone marrow to make a dressing for his hair. Women, of course, had been using arsenic as a cosmetic for centuries, either topically as a paste or ointment like Young John, or swallowing it in sub-toxic doses, but by the nineteenth century doctors were prescribing it for practically everything from asthma to typhus, malaria, period pain, worms, anaemia, syphilis, neuralgia and as a general pick-me-up. At a Westminster Medical Society meeting in 1829, one veteran member described arsenic as one of the most powerful of tonics. After taking it people felt 'an unusual excitement of the system' and their whole frame was 'wound up after the manner of a musical instrument over-strung'.

The favourite mixture was a proprietary brand called Fowler's, developed to treat what was known as ague (fever and chills, usually related to malaria) by a Staffordshire doctor of that name in 1786. A one per cent solution of potassium arsenite, Fowler's was still being prescribed in the 1930s, yet 100 years earlier the eminent surgeon Sir Astley Cooper had told his students: 'The bad effects which this medicine produces ... often lead us to regret that we should have employed it at all.' Indeed, patients died of the 'cure' on a regular basis: a post-mortem on a sixty-three-year-old man with cancer of the tongue who was given arsenic for ten days concluded: 'The remedy rather than the disease had terminated existence'. And after recommending Fowler's as a powerful tonic and effective treatment for ague, intermittent headache, leprosy and 'other obstinate cutaneous diseases', an 1830s pharmacopoeia mentioned in passing that it was contraindicated in pregnancy 'as it is apt to produce abortion by destroying the life of the foetus'.

Another pharmacopoeia, co-authored by the Edinburgh toxicologist Robert Christison, noted that many doctors were following the practice of a Dr Blackadder in using Fowler's to treat hospital-acquired gangrene, applying the solution to the affected site at frequent intervals until the surface was covered with dead skin.

As early as 1809, doubts were also being expressed about

arsenic's efficacy. 'It certainly produces at times a salutary change in the appearance of the sore,' said *The London Medical Dictionary*, before adding: 'We have had reason to regret this change is not permanent.'

Even so, trained medical men and quacks alike continued to use arsenic in various topical preparations. 'Miss Plunketnet's receipt' consisted of arsenic, sulphur and the leaves of the greater crowfoot and the lesser crowfoot, mashed to a paste, formed into balls and dried in the sun. 'Before use, beat with an egg yolk and apply on a pig's bladder,' the instructions ran. Monsieur Febure, on the other hand, favoured a liquid wash made up of arsenic, extract of *Cicuta* (water hemlock, also highly poisonous), Goulard's extract (a lead-based lotion) and laudanum, to be applied every morning.

Young John's account of how he came to catch such a socially embarrassing disease put Middle John into further bad light. 'My father brought home the complaint when he cohabited with a female named Warren and lived apart from my mother,' he told the court. In fact, on 21 August 1823, Hannah Warren had given birth to a daughter, Mary-Ann. The father was named as one John Bodle, occupation, farmer. That summer of 1823 was clearly eventful for Middle John, for it was the same year that he stood accused of cutting hop bines.

On the evening of 2 November, the day the family were taken ill, Young John continued, Middle John had come home and said they were all bad at his grandfather's. 'My father said they expect they are all poisoned. He then told my mother to put on her bonnet and go up there as they were all vomiting, and grandfather was as sick as the devil.'

He had never talked about wanting his father and grandfather dead or said he wouldn't mind poisoning anyone who
offended him. He was on good terms with his mother but
he did remember a conversation about heating something
up over the fire – it was indeed marrow for his hair. And he
hadn't run away to London on the morning his grandfather
died; he had a letter, dated 24 October, from a cousin in
London to his mother about some money, which ended:
'Tell John to come to town on Tuesday week as it will suit
better.' Middle John had of course earlier told the court he
had no idea where his son had gone or why.

Carttar then asked him: 'Will you swear whether when
you filled the kettle you put anything in it?'

'I am quite sure I did not.'

'Will you positively swear you did not?'

'I swear positively I did not.'

'Did you ever carry any arsenic about you?'

'Never.'

Carttar wanted to hear from the doctors about the danger
of using arsenic on the skin, quoting cases from Robert
Christison's *Treatise*, one detailing a young man who had
used an arsenic preparation for an 'itching eruption' and
was ill in bed the next day, and the other concerning a
woman who again had tried to cure 'an inveterate itch' with
arsenic lotion. According to Christison, she had died from
an attack of erysipelas, or scarlet fever, as a direct result
of the treatment. Presumably the coroner was wondering
whether Young John's account of the quantity of arsenic he
got through was believable. Butler and Francis Bossey both
said a cold arsenic solution would certainly dry up itchy
pimples. They believed Young John's mixture of 6 ounces

of water and half an ounce of arsenic was dangerous but wouldn't go so far as to say it would definitely have made him ill. The ointment would be 'less corrosive' because of the lard, Bossey thought.

With the evidence now in, the coroner began to sum up. The jury was being asked to enquire when, how and by what means George Bodle had come by his death, he told them. The question of when had been proved by several witnesses; how was proved by the evidence of the medical gentlemen, but by whom or by what means was not so satisfactorily explained.

Death was caused by a small quantity of arsenic and it was evident that the poison had been in the family's coffee. Mr Evans, the proprietor of the Woolwich chemist's shop, had traced the possession of arsenic to the prisoner, which he said he had purchased to destroy rats – given the complaint he was suffering from, however, it was not extraordinary that he should conceal the purpose for which he required it.

Turning to Constable Morris, the coroner said his evidence was 'so contradictory to the truth that he should entirely put it out of the question'. In fact, if there were grounds for charging the prisoner with having administered the poison, Morris had 'completely knocked the case on the head by exhibiting the packet of poison he found in the prisoner's box to the company of several public houses where it was handed about from one to another'. The prisoner's lawyer had argued that the arsenic missing out of the packet was lost by a drunken Morris who had allowed it to be shown about, fingered and 'taken up in the same manner that a person would take a pinch of snuff'.

When it came to Mary Higgins, Carttar was inclined to

give her some credence: he didn't think that she had set out to mislead the court. Mrs Bodle had contradicted Higgins but the jury had to remember that this was a mother being asked to testify against her son; it was only natural that she should 'lean towards him'. Middle John was a different matter. His evidence was full of contradictions, did not tally with that of any of the other witnesses and contained so much 'important matter affecting the life of his son' that the coroner said he would leave the jury to make up their own minds. When he saw a father 'so anxious to saddle his son with the commission of such a crime', however, he could not avoid remarking that he considered it exceedingly unnatural and was quite at a loss to account for it. Nor was he convinced by Young John's story of the beggar man coming to the gate on the morning of the poisoning.

In reaching their verdict, the jury need not consider whether George Bodle was the intended victim: 'If A intended to shoot his brother but accidentally killed his sister it was murder as much as if he killed the person he intended.' In cases of poisoning, it was enough to find the person who had administered it.

Carttar then sent the jury out, reminding them that if they went against the prisoner, he would be sent for trial where he would be prejudiced by their verdict.

The jury trooped back after only half an hour. 'The verdict of the jury is that John Bodle, the younger, the prisoner, is guilty of wilful murder against George Bodle,' announced the foreman, John Ward.

Carttar, who if trying to appear impartial was not being entirely successful, told Young John: 'It falls to my lot to announce to you the verdict ... is wilful murder against you.

It will be my duty therefore to commit you to Maidstone gaol for trial and I trust that better success will attend you there than it has here.' The young man showed no reaction, just shook hands with some of his friends before being led out.

There then followed another row about the cost of the prosecution that ended with the coroner threatening to send one of the executors of George Bodle's will, the undertaker Henry Mason, to join Young John in Maidstone prison unless he agreed to pay. It was 10.30 p.m. before the proceedings finally broke up.

Young John's composure did not last long. His lawyer Mr Colquhoun found him in the little room where he had been led, quite hysterical, tears streaming down his face and in a state of collapse. Two constables were holding him upright. John Butler was sent for but there was little the surgeon could do except administer brandy and try to console him. Butler and Colquhoun both reminded him that the verdict did not mean that he was guilty and that he would still have the chance to prove his innocence at the forthcoming trial. Young John calmed down a little at this but continued to weep and insist he was innocent. When he began to cry 'Oh my poor mother' over and over, Mr Butler set off in his phaeton to fetch her. 'On beholding her the prisoner burst into a flood of tears, took hold of both her hands and, with his head resting on them, wept bitterly. The interview was of short duration but most affecting,' *The Times* said.

Young John spent the night locked up in Plumstead cage, the little prison cell next to the poorhouse on the main road, watched over by two constables. The next morning his mother, sister and other female members of the family came to say goodbye. This was also the day of George Bodle's funeral at St Nicholas church.

As the funeral procession made its dignified journey to the ceremony, there passed by on the other side of the road a large locked van, the horses pulling it clipping along at a somewhat smarter pace than the hearse. Young John was on his way to Maidstone jail.

After the service, eighteen members of the family dined at Bodle's farmhouse. 'The father of the prisoner did not appear much disturbed at the melancholy situation of his son but partook of dinner and was apparently the least concerned or affected of the whole party,' noted *The Times*.

One interesting new fact emerged before Young John's trial began. Another inspection of the farmhouse cupboard revealed that the morning ceremony of the keys, with the old farmer coming down to unlock the cupboard and measure out the coffee, had been a charade. The lock on the cupboard door was broken; anyone with access to the parlour could get at the coffee, while only a sliding panel separated the coffee cupboard from the sugar cupboard, which was kept unlocked all day. And after the coroner had confiscated Baxter's set of keys and shown them to the jury, they were found to be 'of a very common description'.

A week later, on 23 November, PC James Morris was suspended from duty as a prelude to being dismissed for drunkenness and incompetence. His seems not to have been an isolated case. In explaining the decision to dispense with Morris's services, another magistrate, Adam Young, reported that 'serious complaints had on former occasions been preferred against the constables of Woolwich for suffering persons of notoriously bad character to escape'. He called upon the high constable to appoint a replacement for Morris and trusted that a similar circumstance would not occur again.

~~ 14 ~~

From the Very Brink of Eternity

As November gave way to December in west Kent, the mild weather continued, sending nature awry. The local paper was reporting unseasonal phenomena such as gooseberry bushes and pear trees coming into fruit, white violets starring the banks of the roads and summer flies swarming over the fields.

The 1833 Kent Winter Assizes opened for business at Maidstone on Tuesday, 10 December, before Sir Stephen Gaselee, justice of His Majesty's Court of Common Pleas, and Sir John Vaughan, baron of His Majesty's Court of Exchequer. The grand jury, whose job was to decide whether the accused had a case to answer, was presided over by Lord Brecknock and made up of local worthies, including the Reverend Baden-Powell, a professor of geometry and father of the founder of the Scout movement. Powell had known George Bodle; he had been the vicar at St Nicholas church in Plumstead in the early 1820s when the farmer was churchwarden there. The petit or petty jury – tasked with hearing the cases the grand jury put before them – included chemists, carpenters, farmers, butchers, grocers, a cordwainer, an auctioneer and a smattering of 'gentlemen'.

The ceremonial opening of the proceedings had taken place the day before, after which their lordships attended church. One hundred and thirty-four names were on the calendar, including two people accused of murder, four of rape and four of attempted rape. Studying the list, Gaselee remarked that assaults on children of tender years had greatly increased recently and it would be necessary therefore 'to introduce a check by very severe punishment'.

Among the cases awaiting the jury's attention was that of two soldiers, George Cropper, twenty-seven, and eighteen-year-old Charles Pike, charged with 'an abominable offence'. 'Cropper feloniously, wickedly and diabolically and against the order of nature, did carnally know the said Charles Pike,' the indictment ran. He 'perpetuated the detestable, horrid and abominable crime amongst Christians not to be named … to the great displeasure of Almighty God and to the great scandal of all humankind'. Pike was acquitted but Cropper, seen as his seducer, was found guilty and hanged. Elliott and Richard Kett, committed by the Woolwich magistrates William Stace and the Reverend Dr Samuel Watson for stealing wheat, a sack and a barn cloth at Plumstead, would be transported for fourteen years, while seventeen-year-old David Crump was to be whipped and imprisoned with hard labour for one calendar month for stealing a hop pole, value 2d.

The seventy-one-year-old judge Sir Stephen Gaselee, a leading member of the Royal Humane Society, was to be satirised three years later in Charles Dickens' first best-seller, *The Pickwick Papers*. Gaselee was cast as Mr Justice Stareleigh, 'a most particularly short man, and so fat that he seemed all face and waistcoat. He rolled in upon two little turned legs, and having bobbed gravely to the bar, who bobbed gravely to him, put his little legs underneath his table, and his little three-cornered hat upon it; and when Mr Justice Stareleigh had done this, all you could see of him was two queer little eyes, one broad pink face, and somewhere about half of a big and very comical-looking wig.'

Dickens' jibes weren't confined to Gaselee's appearance: 'Serjeant Buzfuz ... here paused for breath. The silence awoke Mr Justice Stareleigh, who immediately wrote down something with a pen without any ink in it, and looked unusually profound to impress the jury with the belief that he always thought most deeply with his eyes shut.' At the time of the Bodle trial, the twenty-one-year-old Dickens was working as a shorthand writer in the London courts and so had the chance to study Gaselee at close quarters. It was said that the judge, who was in fact generally regarded by fellow lawyers as conscientious and well informed, was very upset by the parody and never entirely recovered for the remaining three years of his life.

After Gaselee's opening remarks, Lord Brecknock warned the assembled petty jury that much would be demanded of them. 'The number of prisoners is more than could reasonably be expected and many of the crimes [are] of a very heinous description,' he said, adding: 'There is one case of a very serious nature, a charge of murder of the worst possible description. The name of the accused is John Bodle and he is accused of poisoning his grandfather. When I have said this I have said enough to induce you to pay to this case the most serious consideration. You will first consider whether the deceased really died from taking poison and then whether the accused administered it.'

'The court did not sit until near 12 o'clock', remarked *The Times*, 'and nothing occurred before our parcel [of news reports] left which was at all worthy of notice.' The trial of John Bodle was a different matter and the paper's coverage was extensive. Also crowding into court were the reporters from other news outlets: the national dailies such as the *Morning Chronicle* and the *Morning Post*, some of the larger regional papers and the freelance 'stringers' who supplied material to papers all over the country. *The Times* sniffily referred to the latter as 'those inventive geniuses commonly called penny-a-liners'. (Dickens began his journalistic career at the age of fourteen by hawking penny-a-line paragraphs on accidents, fires and crime around the London newspaper offices.) But despite his lofty tone, the editor of *The Times* was not above using penny-a-liners himself, as was revealed in 1837, when his paper published a story supplied by one of their number claiming that the discovery of a human head in an Ealing country house and the disappearance of a young woman 'of considerable personal attractions' were being linked in a murder inquiry. The head turned out to be part of a skull used in an anatomy class, the police were not trying to trace a young woman, attractive or otherwise, and the entire piece was 'a fabrication', *The Times* was forced to admit.

Inauspiciously Young John's case was listed for Friday the 13th. Gaselee arrived at 8 a.m. for the start of the hearing at 8.30. As the courthouse doors swung open, the crowds, some of whom had been waiting for hours in the dark to secure a good place, rushed into the building, and the galleries and benches were packed with those who were, in the words of *The Times*, 'exhibiting a most intense desire to catch every word of the evidence'. Many of them were well-dressed women.

The purpose-built Maidstone sessions house, built at a cost of £29,000, opened in 1826, replacing the town hall that had doubled up as a home for the assizes. The old building was notoriously inconvenient and uncomfortable. In the bitter winter of 1823 Mr Justice Park called it a disgrace to the county, saying that the county had lavished 'enormous sums erecting a splendid gaol' but had allowed the judges of the land to sit in a mean and insignificant court 'almost at the peril of their lives'.

The new sessions house backed on to the jail yard for the easy transfer of prisoners. The two-storey court consisted of a main block with a winged colonnade to each side.

The main doorway led into an entrance hall, with a grand staircase to the upper floor immediately ahead, the crown court to the right and the civil court, known as the *nisi prius*, to the left. A row of rooms for the crown court attorneys, the clerk of indictments and the witnesses led off the entrance hall. Behind them the main crown courtroom consisted of a raised area with the judge's seat dominant at the back, flanked by the high sheriff to his right and magistrates' benches on either side.

Counsel sat in a semicircle facing the judge, with the jury behind them to their right and the witness box to their left. In between the jury bench and the witness box was the dock, directly facing the judge, and behind the prisoner, around the outer edge of the semicircle, were seats reserved for the crier, the prison governor, the bailiff and the under-sheriff.

The back half of the chamber consisted of blocks of seats for the public, the witnesses and the jurymen waiting to be called to hear a case. Two staircases led down to the basement, one from the dock so that the prisoner could be led through the yard to and from the jail, and one from the magistrates' benches to the *nisi prius* court via the basement. The basement also had lock-ups for male and female prisoners waiting to be summoned.

Young John was brought from his cell across the prison yard and into the courthouse to face Mr Justice Gaselee. He had abandoned the fashionable clothes he'd worn at the inquest and was now dressed 'very genteelly', one reporter thought, in deep mourning. The man from the *Standard* found it hard to imagine him as a callous killer, describing his demeanour as 'exceedingly mild', but the reporters disagreed about how he had coped with a month in Maidstone jail and the prospect of a death sentence. To the *Morning Post* he appeared fit and well, but *The Times* opined that the young man now looked very different from the one who had faced the coroner's jury at the Plume of Feathers four weeks earlier: 'His appearance showed that he had been considerably affected in health since the inquest.'

Even if *The Times* was right, Young John was still fortunate: a few years earlier his situation would have been far worse. The old Maidstone jail and house of correction had been filthy, overcrowded and foul smelling, with thick wooden boards over the windows shutting out both air and light. There were no covers on the beds, prisoners were frequently put in irons or whipped, while the louse-spread disease, typhus, was a constant threat. In the late eighteenth century, twenty-two prisoners had died of the appropriately named jail fever. The most elaborate and expensive county jail of its time, the new Maidstone prison opened in 1819 to house 450 prisoners. It was designed along the lines suggested by the reformer John Howard, with the prisoners in single cells, and separated into males and females, debtors, convicted prisoners, those like Young John awaiting trial, and juveniles.

Young John Bodle was one of 100 male prisoners awaiting trial at the time. Held in the common jail, as opposed to the house of correction for those already sentenced, the inmates were subdivided again according to their previous character, their conduct in the jail and the seriousness of the charges. The charges did not come any more serious than that facing Young John but, innocent until proved guilty, he was entitled to better treatment than the convicts.

Because he was awaiting trial, Young John was not forced to survive on prison food. His family and friends could send in provisions, although wine was permitted only on the orders of the surgeon. His small cell contained an iron bedstead with a straw mattress, two blankets and a rug, but he was also allowed 'greater indulgence in bedding, linen or other necessaries' and he could wear his own clothes instead of the coarse woollen jacket, waistcoat and trousers of the men in the house of correction.

Prisoners were locked in at sunset and lights-out was at 10 p.m. Those sentenced to hard labour began work at dawn in winter and stopped half an hour before sunset, with half an hour for breakfast and an hour for dinner. When the jail first opened the main work was spinning, but in 1824 treadmills were introduced: large, hollow wooden cylinders on an iron frame, with steps 7.5 inches apart. The prisoners, male and female, gripped the handrails on either side and climbed a never-ending staircase, powering a mill that ground corn and pumped water. Together the two tread-mills at Maidstone could accommodate over two hundred prisoners at a time, but one served no useful purpose, being purely for punishment. The prisoners plodded on in silence, separated from each other by partitions, for six hours a day in three-hour shifts, with fifteen minutes on and five min-utes off, climbing the equivalent of nearly nine thousand feet. Awaiting trial, Young John was spared this demeaning experience, and he was also allowed extra visiting hours so that he could consult with his lawyer and prepare his case.

In court, Young John was described as being of no occu-pation rather than as 'gentleman', and the clerk of the arraigns read the indictment. The 'good and lawful men' who had made up the coroner's jury said that John Bodle the younger, 'not having the fear of God before his eyes and being moved and seduced by the instigation of the Devil and of his malice aforethought', had with poison feloni-ously and wilfully killed and murdered George Bodle on the second day of November. To do this, John Bodle had privately and secretly conveyed 'a large quantity of white arsenic, to wit two drachms', into the kitchen of the dwell-ing house of the said George Bodle, and on the same day, the same white arsenic with a certain quantity of water in a certain kettle in the said kitchen, did put, mix and mingle; the said John Bodle well knowing the said white arsenic to be a deadly poison.

George Bodle used a great quantity of the water with which the white arsenic was mixed and mingled by John Bodle, the younger, to make a certain liquor called coffee, the indictment went on. And George Bodle did take, drink and swallow down a large quantity of the white arsenic, 'he, the said George Bodle, not knowing there was any white arsenic or other poisonous or hurtful ingredient mixed and mingled with the said water, by means whereof, the said George Bodle of the poison there became sick and distempered in his body and the said George Bodle of this poison aforesaid so by him taken, drunk and swallowed down and of the sickness and distemper occasioned thereby from the said 2nd day of November did languish and languishing did live, on which 5th day of November of the poison and of the sickness and distemper occasioned thereby did die'.

The said John Bodle the younger pleaded not guilty.

A huge number of witnesses had been lined up: twenty-six for the prosecution and twenty-five for the defence, but the marathon Plume of Feathers inquiry had been something of a dress rehearsal and, by whistling through the evidence, this court dispensed with the case in less than two days. The prosecution lawyer, John Adolphus, who three years earlier had represented the quack John St John Long in the Catherine Cashin case, began by telling the jury to dismiss 'all the prejudices from their minds that their horror of the offence might inspire', but at the same time to dismiss any false compassion, 'for both false prejudice and false compassion are equally destructive to justice'.

Mr Adolphus then said that he wished to call Ann Bodle first. He had been told that she was exceedingly ill and he wanted to get her ordeal over with so that she could go home. Mr Clarkson said he had no wish to cross-examine the old woman and she was allowed to go back to the Star

Inn where she was staying. The witnesses from the inquest, including the doctors Butler, Solly and Bossey and James Marsh, then trooped into the box to repeat what they had previously told the coroner. Butler testified that the post-mortem had revealed 'recent mischief in the stomach'.

Constable Morris, now suspended from duty, provided a little light relief by making a last-ditch and, as it turned out, unsuccessful attempt to save his job. The reason for his suspension had never been explained to him, he told the court, but he had heard that it was in consequence of his having been intoxicated. He had had some rum at Mary Andrews' coffee house when he arrested Young John, he admitted, but it was very little – only part of a glass – and Mrs Andrews had been to blame. 'I did not pay for the rum, nor did I propose it should be sent for,' he insisted. Before he and the prisoner boarded the coach at the Cross Keys in Gracechurch Street, they did have a glass of ale apiece but just one brandy and water between them, and when they called at the Mortar Inn after their appearance before the coroner, it was only for a glass of peppermint. 'I was asked to drink out of some half-and-half,' Morris said. 'I put my lips to it but did not drink any. That I swear.' The constable confessed to having visited three pubs the following day and having spent seven hours in one of them, but insisted that in all that time he had drunk nothing but a single medicinal glass of ale with some ginger in it 'because I had a pain in my stomach'. He did show the arsenic to Mr Osborne but Osborne merely put his finger in it.

Then Sophia Taylor was called. She denied that Young John had ever filled the kettle before the day the family fell ill or that she had instigated his early-morning visits, but she told his lawyer, Mr Clarkson: 'He is a very nice young man and I know nothing against him.'

Clarkson then changed tack and asked whether she had seen Middle John. Yes, she had seen him the previous night and again that morning at the Star Inn where she and Middle John and Catherine were staying. He appeared very well, Sophia volunteered. But Mr Clarkson had no interest in Middle John's state of health. 'Do you not know that he is a material witness?' he asked. Mr Adolphus was quickly on his feet to object. He had no plans to call Middle John to testify. He only ever called a witness if that person's evidence affected the case. He had read through Middle John's deposition and found 'not one single syllable of evidence in it from beginning to end'. He could not and would not call him.

This was a crucial matter for the defence: Clarkson and his colleague William Bodkin saw Middle John's character in general and his behaviour over his father's death in particular as working hugely in their client's favour. They desperately needed Adolphus to call Middle John as a prosecution witness so that they could cross-examine him: it would then be a simple task to expose their client's chief accuser as a wastrel, a liar and perhaps much worse. Middle John's name appeared on the list of prosecution witnesses on the back of the indictment, and the grand jury had had his statement before them when they made their decision to send the case for trial. Therefore, Clarkson told the judge, the prosecution was obliged to call him.

Usually in such a case the witness's deposition was read out in court so that he could be cross-examined on it, Gaselee replied. 'But if there be any particular reason for keeping it back, the learned counsel for the prosecution must use his own discretion.' Mr Adolphus did indeed have a particular reason for keeping it back, though it was not one he was prepared to share with the jury. He was every bit as anxious not to call Middle John as the defence was desperate to have him, and for the same reasons. The Crown appeared to have wrong-footed themselves by putting Middle John's name on the witness list. 'I understand from my learned friend that he [Middle John] is being studiously kept out of the way,' Mr Clarkson told Gaselee, adding that the prosecution ought to be conducted fairly.

John Adolphus then produced a fine piece of theatre: 'I have been told that I sedulously keep back a witness and that I have not conducted the case fairly,' he announced. 'I hope I shall never attempt to oppress any man in the conduct of a prosecution ... and I call upon your Lordship to say whether in the course of my practice I have ever wished to conduct a case in a manner that did not become a man of humanity, a gentleman and, as far as was in my power, a lawyer. I therefore repel that observation with indignation and scorn.'

Gaselee ignored the histrionics, merely remarking that he wouldn't go as far as Mr Adolphus in saying that there was nothing material in Middle John's deposition because he hadn't read it in any depth, but he did know that many of the depositions in this case were not evidence, one way or the other.

'It is well known, my lord, that John Bodle was the first man who went before the magistrates and laid information against his son and I fear I have no evidence for the prisoner unless the fair and ordinary practice is adopted,' Clarkson said.

'I always recommended the counsel for the prosecution to call the witness unless there was any objection against it,' Gaselee repeated.

'I would not require to be asked; I would do it, my lord, if I thought the deposition was at all evidence,' Adolphus told him, 'but I will neither be induced by caresses or observations to act contrary to what is my judgement and my duty.'

'Oh I will not caress you,' Clarkson promised him. Gaselee said he would discuss the question with his fellow judge and announce his decision later. Everyone then calmed down and the trial continued.

Mary Higgins, who was still living in the poorhouse where the Woolwich magistrates had sent her more than a month before, stuck to her story about what was said at the cottage and denied having any grudge against Young John. Then Betsy Smith, Sophia's fellow servant and Ann Bodle's granddaughter, was called. She gave her evidence in sign language through an interpreter. Although she added nothing of any weight to the case, she clearly found favour with the gentlemen of the press. 'She is a good-looking girl and gave her answers by her fingers with great readiness and self-possession,' said *The Times*. As she left the witness box she curtsied to the jury.

Mr Clarkson then returned to the subject of Middle John, throwing around a ragbag of gossip and innuendo about his character. Sophia Taylor and Judith Lear were both asked about a robbery at the Bodle farmhouse the previous spring. Hadn't Middle John been found at the scene, and weren't there rumours that he was the thief and didn't his father forbid him from entering the house for a while because of it? They said they didn't know.

The lawyer then reeled off the names of various local women. Did Sophia know anyone called Hodges or Shears or Warren or Warwickshire? She knew none of them, she told the court. Judith Lear admitted knowing a woman called Hodges in Plumstead who 'passed for a married woman', and another called Stevenson in a village called Shoulder-of-Mutton-Green. She could not swear that Middle John had been living with the former and had a family with her but she believed he had. As for the woman Stevenson, she had met her only recently and knew nothing of her children.

A few months before the trial the parish had paid a Matilda Hodges 6 shillings out of the relief fund 'to go away for nine months'. And a certain William Hodges had given evidence for the prosecution at Middle John's trial for cutting hop bines.

The case was finally adjourned at nine o'clock that night, and when the proceedings resumed at 8.30 the next morning, the courtroom was packed once again. The jury filed back in, having spent an uncomfortable night locked in the courthouse, sleeping on mattresses. Taking his seat at nine o'clock Mr Justice Gaselee announced his decision over the hard-fought issue of calling Middle John. He had consulted his learned brother Sir John Vaughan and they were of the opinion that while a prosecutor was not compelled to call a witness whose name was on the back of the indictment, the judge was allowed to use his discretion. As the present case was serious, he felt bound to allow the witness to be put into the box.

Thomas Clarkson had got his way, and he proceeded to make the most of it. He began by asking Middle John why he had informed against his son. Because of what his father had said on his deathbed and through a sense of justice, he said. The lawyer then turned to the will. 'I believe I am entitled to the bulk of my father's property after the death of my mother,' Middle John told him. This was not correct, of course, but Middle John told the court he had not known about the new will that gave land to the Baxters. This was probably true. There was no reason for him to lie about thinking he was due to inherit more than he was – quite the reverse, in fact. The implication was that Middle John had stood to benefit considerably more from the old will and that neither George nor Samuel Baxter had told him that he had been partially disinherited a week before George died. Baxter, however, appeared to have told Mr Clarkson something different, for the lawyer then put it to Middle John that the pair had already discussed the matter.

Middle John was next taken through his own movements the day that George Bodle was poisoned. He had got up later than usual, at seven rather than six. At about eight, he had gone across the common to see to Samuel Baxter's sheep and had stayed there until noon. Clarkson then asked him about an incident involving a couple called Jacobs that was said to have happened the morning the family fell ill. Middle John said he knew William Jacobs, the tailor who clothed the inmates of the Plumstead poorhouse and who lived on the common. He couldn't remember whether he had gone to Jacobs' house at ten that morning but if he had, he was certain he hadn't told Jacobs there was a 'devil of a job happened at my father's'. Pressed, he insisted: 'I never said that they had been taking their usual diet for break-fast. I did not tell Jacobs that they had the coffee in large quantities and kept it in bottles. I did not say that the quantity they had was out or nearly out.' This was sensational stuff for, if true, Middle John had known the full details of the poisoning only two hours after it happened, rather than first learning of it eight hours later when he went to collect his wages, as he claimed. When the judge refused to hear William and Frances Jacobs' evidence (he had earlier ruled that witnesses could not be called to contradict Middle John), Clarkson asked for it to be recorded that they were present in court, ready to testify.

Clarkson now turned once again to Middle John's character, repeating the gossip he had put to Judith Lear and Sophia Taylor and adding more besides. During a litany of accusations, the judge repeatedly interrupted to remind the witness that he need not answer any of the questions if he didn't want to. He did not want to, although his outraged replies served Mr Clarkson's purpose well enough: 'I object to answering whether I have ever been charged with a felony or that I have ever been in Maidstone gaol or that I have families living by three different women besides my wife or that I was ever charged with attempting to cut my wife's throat.' Mr Justice Gaselee then stepped in again to warn Clarkson off once and for all, at which point the lawyer said he had several more questions of the same kind to put to the witness but would not press them, thus leaving in the jury's minds the picture of a further troupe of skeletons jangling in the cupboard.

Samuel Baxter came in next, to give character references for the prisoner. Baxter seemed particularly agitated and he wept as he told the court he had known the young man since he was born. 'I never saw anything but what was right of him in my life or else I would not have suffered him to come to my family,' he said. Mason too had known 'the lad at the bar' from his infancy: 'I never knew anything against him in my life. He has been respected and beloved generally by his relatives and acquaintances.'

At this point the judge told the lad at the bar that it was time for him to answer the charges.

Young John stated: 'I am perfectly innocent and my defence is embodied in this paper,' and handed a document to a court official. The court then sat in stillness as the disengaged voice of a clerk read what appeared to be the impassioned words of a young man pleading for his life. The statement was, in fact, a beautifully crafted piece of work by one of his legal team, William Bodkin. The *Chelmsford Chronicle* described it as a masterpiece of composition, judgement and sound reasoning.

The evidence at the inquest about helping regularly with chores at the farmhouse and using arsenic to treat the itch was repeated. Then Young John turned to the matter of his trip to Clerkenwell. 'Gentlemen, of all circumstances which justify suspicion and insure condemnation, the flight of a party accused of a criminal offence is one of the most powerful and conclusive. It is almost always considered as equivalent to a confession of guilt.' But he had already arranged to visit his sister, 'to whose house I proceeded, being at the time unaware that my poor grandfather was so near his end'. If he was guilty, why hadn't he run away earlier? 'His sister Mary Andrews, so distraught at seeing her brother in the dock that the judge allowed her to give her evidence sitting down, then produced the letter that Young John had mentioned at the inquest from Henry Perks' sister to Catherine

Bodle about the purchase of some merino wool that ended: 'Tell John to come to town on Tuesday week [5 November] as it will suit better'.

As for Mary Higgins' claim to have overheard him planning to kill his grandfather: 'Gentlemen, I solemnly declare that no expression of the kind ever passed my lips – no such thought ever entered into my imagination. This part of the case rests entirely upon the evidence of my unnatural father and of his servant Mary Higgins.'

The rest of the statement drew attention to the strange conduct of the man the defence now claimed was the true murderer. 'You will recollect that my own father is the first person to accuse me of this dreadful crime. He it is who goes to the magistrate and puts this charge in motion against me. He it is who, after an interview with Mary Higgins at five o'clock in the morning, produces her as a witness to insure my condemnation by swearing to a conversation that never occurred, and in the relation of which they falter and equivocate in a manner you have heard. Gentlemen, what honest or proper reason can be assigned for that remarkable interview which took place in the dark at so early an hour between them? What was its object but to concert that evidence which it was hoped – but I trust vainly – would insure my condemnation?'

And the motive? Even if he knew his son to be guilty, would not a father normally abstain from being the first to prefer an accusation that might bring his own son to a speedy and ignominious death? Middle John, he was aggrieved to say, had not set a worthy example. It had been his children's lot to see him imprisoned for malicious injury, and guilty of profligacy of all kinds, 'but still the voice of nature cannot be so absolutely dead within him as not to have left him the ordinary regard which even the brute creation exhibit towards their offspring. What then is the great, the overwhelming, motive that has made him thrust himself forward uncalled for, unsolicited, to take away my life?'

The answer was clear. 'You will have observed in every stage of this proceeding, an extraordinary anxiety on the part of my father to show his own innocence of this crime and my guilt. Before suspicion attached to anybody you find him the first to ask who filled the kettle that morning … You find him proclaiming the fact of his not having been at my grandfather's house that morning; and you find him lying in bed much later than usual as if on purpose to be ready to show where he then was. But, gentlemen, how is it shown that he was not at my grandfather's the day before, when arsenic might so easily have been put into the coffee jar, and the more easily because Sophia Taylor was out the whole of that day?' And in his conversation with the Jacobs on the morning of the poisoning, Middle John had given himself clean away by showing that he knew the coffee that had been used for breakfast was the last in the jar. How could he have known this?

The prisoner then thanked the jury for their patience in listening to his 'afflicting tale' and finished by reminding them of the torment that would be theirs if they sent him to his death only to find later that he was innocent. 'I beseech you to remember that I am addressing you, as it were, from the very brink of eternity and that, if I fail to convince you of my innocence a very few hours will remove me from among the number of the living …' He then 'with humble confidence' left his life in the hands of the jury and prayed that God 'to whom all hearts are open, and from whom no secrets are hidden, may so watch over and influence your deliberations, that you may at last arrive at a just ... at a humane and conscientious decision'.

Hard as this was to beat for dramatic effect, the defence chose to follow Young John's deposition up with more character witnesses, each seemingly vying to see who could heap the greatest praise on Young John. While Mr Terry, the local tax collector, had always known him to be an honest, steady, inoffensive young man, and David Rice, landlord of the Plume of Feathers, believed he had always borne a good character, the Cleeve family's contribution bordered on eulogy. Edward Cleeve said the prisoner had always borne the best of characters for humanity and kindness of disposition; Thomas Cleeve said he could not be excelled for tender feelings, humanity and good moral conduct; while Henry Cleeve said his character could not be better and he had never heard him say an ill word about anyone. Farmer Richard Clements went even farther, saying he was the best of creatures who ever came into Plumstead, adding: 'I believe that if he saw a worm lie in his way he would turn out of his road sooner than hurt it.' Others who added their twopenn'orth of praise were the tailor William Jacobs, James Russell, a farmer of Horton Kirby, and John Russell, a Camberwell coal merchant.

Finally two last-minute witnesses were called. Sarah Perks, Henry's sister, who had written to Catherine Bodle asking for Young John to come to Clerkenwell, told the court that for a few months in 1831 she had shared a house in Shoreditch with the prisoner and he was in the habit of using arsenic on his skin there. Another woman, Elizabeth

Brett, had come forward after reading about the case in the newspapers. Her late husband had worked in the Shoreditch shop and Young John had sent him out several times to buy arsenic and hellebore for his skin. She had seen the prisoner mixing arsenic with lard.

That completed the case for the defence. The crime had attracted considerable public notoriety, the judge told the jury, but they must put everything they had previously heard about it out of their minds and concentrate solely on the evidence.

After this warning, Mr Justice Gaselee began his summing up. Hardly had he drawn breath, though, when he was interrupted by the foreman. The jury could save his Lordship the trouble of reading any further; they had already heard enough to come to a verdict. The judge said if that verdict was not against the prisoner, he would trouble them no further. The jury then whispered quickly among themselves and the foreman stood up again: 'My lord, we are quite satisfied and our verdict will not be against the prisoner.'

'Few scenes have been witnessed in a court of justice so exciting as that which took place at the termination of the trial of John Bodle for the horrible crime of which he stood accused,' said the *John Bull* magazine. Young John's friends and most of his family sent up a loud cheer while the rows of 'respectably dressed' women in the public gallery clapped and waved their handkerchiefs. 'The acquittal was triumphant ... The tears of hundreds bore testimony to the sympathy felt for an individual who had during so many hours, stood between life and death.'

The prisoner looked bewildered and seemed not to take

the verdict in but his friends roused him, rushing the dock to grab his hand and clap him on the back. The *John Bull* reporter sensed that the 'humane judge' seemed to share the general feeling and the paper carried a snatch of conversation overheard between Gaselee and the Reverend Dr Watson, the Woolwich magistrate who had set the criminal process in motion and who had sat on the bench beside the judge throughout the trial. 'This verdict I think cannot be found fault with,' remarked Gaselee, to which the magistrate replied: 'A true verdict, my lord, and I think it is your lordship's judgement that the coroner could not have done otherwise than send him to trial?' 'Impossible,' said Gaselee.

'The close of this trial presented one of the most extraordinary scenes ever witnessed in a court of justice,' commented the *Essex Standard*. 'A youth, accused of the murder of his grandfather, charges his own father with the crime and also with that of attempting ... to compass the death of his own unoffending child; and this strange and appalling statement was apparently so far justified that immediately after [it] the prisoner is acquitted, almost by acclamation.'

After the Bodle verdict, the celebrations continued in the streets. Young John was greeted by cheering crowds outside the gates of the courtroom and the throng then proceeded to follow him to the Star Inn, where his parents and grandmother, along with Sophia Taylor and the other witnesses, were staying. A show far more dramatic than anything Madame Palermo and the Infant Prodigy could have staged then took place. The ailing, newly widowed Ann Bodle took her step-grandson into her arms 'with expressions of the warmest affection'. No sooner had this touching scene occurred than the entire jury turned up at the inn, lining up to shake Young John's hand and wish him health and happiness, saying they hoped he would live many years as a useful and honourable member of society. They had already decided by the end of the first day that there was not enough evidence to convict him, but having heard the case for the defence they were now completely assured of his innocence. A coach then bowled up to carry the hero and his party back to Plumstead. When he finally arrived at midnight the news had already reached the village and more well-wishers were waiting to welcome him.

The Sequel of These Proceedings

No one could have been in the slightest doubt that George Bodle had been deliberately poisoned by someone close to him, but the verdict at Maidstone left the authorities at a loss as to how to proceed. Other members of the Bodle clan besides Middle John had both motive and opportunity, as the investigation had shown, and the behaviour of some of them had been decidedly odd, but there was still no hard evidence against anyone. And then of course there were the costs to consider.

'What may be the sequel of these proceedings it is impossible to foretell,' commented the *John Bull*, 'but we cannot close without expressing our hope that as the innocent is acquitted, the guilty may be discovered and brought to justice, for a blacker and more diabolical murder was never perpetrated.' The *Spectator* thought that the matter should not rest there: someone ought to be prosecuted for something, in particular Middle John. 'We noticed at the time the suspicious circumstances that pointed at him,' the magazine remarked. 'Here is a man suffered to go at large who, if the allegation made against him at the trial be true, is a criminal the atrocity of whose guilt is rarely to be paralleled.'

The day after his acquittal, Young John, still dressed in deep mourning, accompanied the Cleeve family to a packed St Nicholas church for a service of thanksgiving for his deliverance. The vicar, the Reverend Henry Shackleton, and his curate Mr Kimber squeezed into the pew to shake his hand and the young man was 'evidently much affected', reported the *Kentish Gazette*.

Young John did not go back to his parents' cottage, of course – living under the same roof as his father would have proved something of a challenge. Instead the young man went to stay with the churchwarden Thomas Cleeve, who had sung his praises at the trial to such good effect. Indeed, such was the sympathy for the young man that some leading lawyers, including his defence counsel Thomas Clarkson, urged him to ask the Court of the King's Bench to quash the coroner's jury's guilty verdict, thus clearing his name completely. Perhaps having had enough of courts and legal arguments for the time being, Young John chose to let the matter drop.

The following May, the parish voted to give Mary Higgins 10s to retrieve her few belongings from the pawnshop and £2s 9d for her fare to Boughton so that she could depart Plumstead for good and live with her father. She disappeared, leaving the village still speculating about what exactly she had done and how much she had known.

Middle John continued on in the cottage alone while the farmland and outbuildings, now in the lifetime possession of his stepmother Ann Bodle, were leased out in lots. Catherine left him and moved into a small rented cottage in the village. Young John meanwhile rented one of the larger cottages for a few months until some of his well-wishers helped him set up another shop, this time in Bishopsgate Street in the City of London, so that he might become the useful and honourable member of society that the jury had hoped.

The publicity over the Bodle case did not end with the trial. Four months later the row over the cost of the meticulous but ultimately fruitless investigation into George Bodle's death, which had led to Charles Carttar threatening to send Henry Mason to prison, erupted once again. By now, though, the matter had reached the attention of Parliament.

On 15 April 1834, a House of Lords committee chaired by the Duke of Richmond summoned John Clark to appear before it. Clark was the deputy clerk of assize to the Home Circuit, the section of the criminal court system that covered the county of Kent. The committee was looking into how the county rates were being spent and had a few questions for Mr Clark. Did he remember the case of John Bodle junior, indicted for murder? He did. Did he recall the judge Mr Justice Gaselee decreeing that the county should pay for the prosecution? He did. In that case, could he explain why the Kent ratepayers had ended up footing a bill of £23 12s for fees to counsel and £27 6s for briefs? Their lordships also had questions about Charles Carttar's bills and Mr Justice Gaselee was also in the frame. There had been some discussion about the number of prosecution witnesses, but in the end the judge had ruled they should all give evidence, Clark recalled, adding: 'I should say that there were no more witnesses than were necessary to make out the case.' Another tetchy exchange followed, about why witnesses in

Kent and Sussex were paid a shilling more in expenses than in Hertfordshire, Essex and Surrey. Clark didn't know; that was just how it was done. Ultimately, though, however irritated the committee might have been, there was nothing they could do about the costs of pursuing Young John Bodle apart from subjecting Mr Clark to an uncomfortable half-hour and making their displeasure clear.

A petty squabble about expenditure at Kent Assizes was not to be the final word on the death of George Bodle, however, although what the *John Bull* magazine referred to as the 'sequel impossible to foretell' would be some time coming.

In the meantime James Marsh's experience in the witness box had sent him back to the lab, pondering a new challenge: to find a definitive way of detecting the smallest speck of arsenic and separating it from the material containing it. What he had not told the courts was how unsatisfactory he had found the standard tests to be: incapable of detecting small traces of the poison, unreliable if the sample contained organic matter such as food, vomit or stomach contents, and, in the case of one experiment, relying on the entirely subjective sense of smell of the analyst.

'Notwithstanding the improved methods that have of late been invented of detecting the presence of small quantities of arsenic in the food, in the contents of the stomach and mixed with various other animal and vegetable matters,' he wrote, 'a process was still wanting for separating it expeditiously and commodiously, and presenting it in a pure unequivocal form for examination by the appropriate tests.' In other words, science needed to come at the question from a different place.

Marsh started to worry away at the problem in any time he could spare from his official duties at the Royal Arsenal and the Royal Military Academy, constructing his own makeshift equipment in the lab. In the autumn of 1836, a paper appeared in the *Edinburgh New Philosophical Journal* under the title 'Account of a method of separating small quantities of arsenic from substances with which it may be mixed'. The author was James Marsh Esq. of the Royal Arsenal, Woolwich.

Marsh's method worked on the basis that when a solution of arsenic in any of its forms comes into contact with hydrogen it reacts to form the gas arsine. So, hydrogen is introduced into the sample – produced by introducing hydrochloric or sulphuric acid followed by zinc – and the metallic arsenic is then recovered from the evolving arsine.

Using this method, the analyst would avoid the risk of a false positive, which was a common problem with the precipitate tests. A yellow precipitate of arsenious sulphide, for example, was often difficult to see in stomach contents, unlike the results of chemical tests for some other substances. (Aspirin in urine gives a purple colour when ferric chloride is introduced, for example. Hospitals once kept a set of tubes showing the different shades, and unless the purple was very pale then a positive result was always easy to see.)

Like the old reduction test, the Marsh test sought instead to reconstitute the elemental arsenic. A jury would be much easier to convince if the scientist could show them the 'mirror' and say: 'This is the arsenic I recovered.' If they were shown an arsenic compound, they might be liable to ask, 'So this is not actually arsenic?', and might then find the subsequent scientific explanation difficult to follow.

For the experiment, Marsh designed a simple apparatus: a 'U'-shaped glass tube open at both ends, with one arm of the tube about five inches long and the other about eight inches. The chemist drops a 1-inch glass rod into the shorter arm followed by a small piece of zinc, which is held in place by the rod. A cork and a stopcock with a fine nozzle are then inserted into the neck of the tube, with the stopcock open.

The sample mixed with diluted sulphuric acid is poured into the long arm of the tube to a level just below the cork. The zinc soon starts to give off bubbles of gas: pure hydrogen if no arsenic is present but arsine if the liquid contains arsenic in any form. The first small amounts are allowed to escape in order to push any air out of the apparatus. The stopcock is then closed and the gas builds up in the shorter arm, driving the fluid up the longer one until the liquid level has dropped below the zinc, after which no more gas is produced.

When the stopcock is opened the gas, under pressure from the fluid, shoots through the nozzle. The chemist then quickly ignites it and holds over it a piece of glass or, better still, a white porcelain dish. Any arsenic present will be deposited as a metal film on the glass. If no arsenic is present, then the glass first mists up with water vapour but then dries quickly and becomes clear.

That was the theory, anyway, and to Marsh's delight it worked. 'I had the satisfaction of finding, on trial, that my anticipations were realised ... I was ... able not only to separate very minute quantities of arsenic from gruel, soup, porter, coffee and other alimentary liquors, but that by continuing the process a sufficient length of time, I could eliminate the whole of the arsenic in the state of arsenuretted hydrogen [arsine], either pure, or, at most, only mixed with an excess of hydrogen.'

The test was extraordinarily sensitive. Marsh worked on dilutions of 1 grain of arsenic in 4 pints of water 'and have obtained therefrom upwards of one hundred distinct metallic arsenical crusts', although patience was needed: 'It required several days before the mixture used ceased to give indication of the presence of arsenic.' Marsh designed a larger apparatus when working with samples of over two pints but the principle was the same. With the small apparatus, he obtained the 'distinct metallic crusts' from one drop of Fowler's solution, which contained just one 120th part of a grain.

He was particularly pleased with the simplicity of the process: anyone could do it, using a do-it-yourself kit even more basic than his own equipment, he believed. 'I may say unequivocally that there is no town or village in which sulphuric acid and zinc cannot be obtained, but every house would furnish to the ingenious experimentalist ample means for his purpose; a two-ounce phial with a cork and piece of tobacco-pipe, or a bladder with the same arrangement fixed to its mouth, might, in cases of extreme necessity, be employed with success, as I have repeatedly done for this purpose.'

The unveiling of Marsh's test was a eureka moment, and hailed as such by the scientific community. Christison called it a beautiful intervention, while the German chemist Liebig described it as 'surpassing the imagination'. The author of *A General System of Toxicology*, Orfila, also took it up with enthusiasm, particularly in his quest to prove that arsenic acted by absorption. He used the Marsh test to extract the poison from human and animal organs, tissues and fluids, working on the bodies of suicide victims as well as his usual tally of hundreds of sacrificial animals. Orfila's discovery meant that analysts could now work with 'pure' samples from the body itself – the liver, spleen, kidneys, muscles, blood and urine, for example – rather than the problematic stomach contents.

The Society of Arts was especially delighted, for this was the project that it had first declared a priority back in 1821, two years before it awarded Marsh the Silver Medal for his portable electromagnetic device.

On 9 April 1836 James Marsh was back before the chemical committee at the elegant Adelphi building off the Strand. 'The candidate gave a very detailed verbal description of his process and illustrated the same by numerous experiments,' the Society reported. This time Michael Faraday, the man who had first set Marsh on his arsenic quest by handing over the Bodle analyses, was not present, but he sent a letter 'highly in favour of Mr Marsh's method and apparatus'. Faraday's views were endorsed by other members of the committee, which now included Alfred Swaine Taylor. The committee 'expressed themselves severally with great and decided approbation of Mr Marsh's investigation, as greatly simplifying the method of detecting arsenic, thus avoiding many sources of errors to which the best of the usual methods are liable'. The Marsh test, they said, was 'worthy of the attention of the Society'.

The test turned out not only to be worthy of the Society's attention but also of its Large Gold Medal. The award ceremony took place on 7 June at the grand Hanover Square Rooms in Mayfair and Marsh went up to receive his award from Vice-Admiral Sir Edward Codrington. By now the public's interest in the latest discoveries in science and manufacturing was such that the Society's house in the Adelphi was unable to cope with the numbers who wanted to attend. The event had been shifted from venue to venue over recent years, and in 1837 the ceremony would be moved again, this time to the King's Theatre in the Haymarket.

Even in his initial excitement, however, Marsh had pointed out two drawbacks to his process. For one thing, the heat needed to burn off the hydrogen could cause the glass to crack and shatter into pieces, while the escaping arsine gas was deadly. 'If the tube, while still warm, be held to the nose, that peculiar odour, somewhat resembling garlic, which is one of the characteristic tests of arsenic, will be perceived,' Marsh explained, adding: 'Considerable caution should be used in smelling it, as every cubic inch contains about a quarter of a grain of arsenic.'

But the risk to the analyst's life and limb was not the only difficulty. Marsh had also warned that the reagent (the zinc) used to create the hydrogen was itself sometimes contaminated with small amounts of arsenic used in the manufacturing process. The analyst could guard against this by first running the test with the zinc and a blank sample. Marsh made it sound so straightforward, yet the problem of impurities in test reagents was not so easily overcome and would nearly ruin the reputation of the foremost forensic toxicologist of his day.

And Marsh's claim that no particular expertise was needed also turned out to be spectacularly wrong. If the sample contained the poison antimony, for example – and antimony, like arsenic, was easy to obtain – this would appear as a dark film on the glass, just like arsenic, and it took some experience to tell them apart. 'Although a practised eye may discern some difference between the crusts, that from antimony being more silvery and metallic,' commented the Bolton chemist Henry Hough Watson, 'yet the line of demarcation is not easily drawn.' Back in 1831 an expert had warned against the dangers of confusing the 'shining sublimates' of two other poisons – mercury and cadmium – with that of arsenic when using the old reduction method.

And while the toxicologists were delighted at Marsh's ability to detect even minute traces of poison, that very sensitivity introduced a new margin for error: the test could equally well pick up minute amounts of arsenic present in cemetery soil, reagents, the vessels used to contain the samples and the environment at a time when arsenic was in common use in paints and pigments.

The following year a Swedish chemist called Jöns Berzelius found a way of measuring the arsenic extracted by the Marsh method, so making the test quantitative as well as qualitative. Berzelius passed the arsine gas through a glass tube that was heated in the middle. The gas was then ignited as it escaped from the reaction solution and the resulting metallic arsenic was deposited inside the glass tube, which was then weighed.

Ironically, it was the case that gave James Marsh an international reputation which demonstrated how difficult the test could be, even in the hands of experienced chemists.

In the autumn of 1840 in the Limousin region of France,

the body of Charles Lafarge was removed from its grave, although 'body' turned out to be a euphemism when his coffin was prised open. Charles had been dead only for a few months but was so badly decomposed that the officials gave up on their lifting equipment and sent instead for a large spoon. The paste that had once been Monsieur Lafarge was then duly scooped up and transported to Tulle in earthenware pots, where a court and chemists were waiting in a makeshift laboratory in a street outside the Palais de Justice.

The Bodle case had engrossed the British public but the trial of Marie Lafarge attracted a global audience. The story had everything. The twenty-four-year-old aristocrat and orphan Marie Fortunée Capelle had been tricked into marrying Charles Lafarge, who had presented himself as a wealthy manufacturer and the owner of a large chateau in southern France. Too late did Marie discover that her husband was on the verge of bankruptcy, the chateau was falling down and she was to share her new home with a grim mother-in-law and an infestation of rodents. Add to this Charles's coarseness and lack of personal attractions – 'his addresses were paid in a manner that shocked her refinement', according to the *New York Times* – and this was clearly not a case of happily ever after.

Just months into the marriage, Charles was plagued by bouts of vomiting and diarrhoea, always after taking food or drink prepared by Marie, who had been buying arsenic, to deal with the rats, she said. When Charles died a few weeks later, a post-mortem and analysis were ordered and the chemists reported finding a small amount of arsenic in his stomach and large amounts in the milk and sugar water that Marie had been giving him.

Madame Lafarge presented a romantic figure in the dock, 'her paleness rendered more striking by the raven blackness of her hair and the deep mourning she wore', said the *New York Times*. The London *Times* reprinted a long article from the Paris-based English-language newspaper *Galignani's Messenger*, quoting another admirer: 'Her conversation ... which is captivating remains ... the character of benevolence ... She always seeks to please and never to eclipse.' And then there was the excellence of her piano-playing, her delightful voice, her competence in more than one science, her reading and translation of Goethe, her fluency in several languages and composing of Italian verse. The plight of the beautiful young widow clearly moved many of the men around her.

Her lawyer began by reading a letter from Mathieu Orfila saying that the precipitate or colour test that the local chemists had carried out was not in itself enough: 'The precipitate itself must be analysed and metallic arsenic found therein,' Orfila wrote. The court duly recruited more experienced chemists, who used 'the most recent methods, particularly Marsh's process' on the samples. They found not the slightest trace of arsenic. Marie Lafarge did not disappoint when the result was announced, clasping her hands, raising her eyes to heaven and then fainting and having to be carried out of court, while her lawyer sat weeping.

This drama turned out to be merely the first act, however. The result of the sample test simply showed that two sets of experts disagreed, the prosecution argued. It was then that the court ordered the exhumation of Charles's body and the extraordinary open-air experiments took place. Once again the chemists found no trace of arsenic, but the prosecution again challenged the results. By now people across several continents were following the toing and froing between the lawyers and the chemists, speculating about what the next set of experiments might reveal.

The judge decided on one last round of tests. Orfila, summoned from Paris by express telegram, arrived in Tulle with two colleagues and began work that same night. For once Orfila declined to put on a public performance. He shut himself away for nearly twenty-four hours, entering the court at five o'clock the following evening. As he rose to speak, a suitably dramatic storm burst over the town and in a darkened courtroom, lit fitfully by lightning, the scientist told his audience that he had four pieces of information to impart. First, he had found metallic arsenic in the body of Charles Lafarge; second, the arsenic did not come from the reagents used in the tests nor from the earth in which the body was buried; third, the arsenic was not that which was normally naturally present in the human body; and fourth, 'it is not impossible to explain the inconsistencies in the evidence of the other experts'. One explanation was that the flame they used may have been too hot, Orfila suggested.

In desperation Marie wrote to the chemist and socialist politician François-Vincent Raspail, also based in Paris. Her lawyers clearly knew that Raspail was no admirer of Orfila – if anyone was qualified, scientifically and temperamentally, to find fault with the toxicologist's results, then this was the man. 'I call to my aid your science and your heart,' Marie wrote. 'Chymical experiments had restored to me a portion of that opinion by which I have been tortured for the last eight months – M. Orfila arrived and I again fell into the abyss.' By now Marie's black hair was streaked with white and she had to be carried into court on a sofa.

Raspail hesitated at first because of his previous skirmishes with Orfila. But realising that it was too late to find anyone else, and after hearing an account of Orfila's experiments which appeared to be 'very defective', he too set off for Tulle. He was, he said, deeply affected by Marie, 'a women devoured by grief', but he tried to hang on to his professional demeanour and appear 'cold as a chymist'. It was all in vain, however; the journey should have taken thirty-six hours but Raspail's carriage broke down and he arrived eight hours too late. The jury, clearly less susceptible to Marie's charm, had already taken just an hour to find her guilty.

Before he returned to the capital, though, Raspail went to inspect Orfila's results: three plates sat in the court registrar's office stained with what the toxicologist had told the court was the arsenic he had recovered from the test samples. Raspail was not impressed. The stains on the first two plates were 'so very small and undefined, and the indications ... so equivocal that I should hesitate to declare them to be arsenical', he said. 'As to the third plate, the stains may be declared to be arsenical but I have many serious observations to make on this point.' One observation was that the chemical Orfila had brought with him from Paris which produced the stain on the third plate had not been tested for purity. When the Limoges chemists raised this, Orfila had offered to withdraw that plate altogether. In that case, the charge of arsenic poisoning fell, as the first two plates could not support such a finding, Raspail said. He also challenged Orfila's estimate that there was half a milligram of arsenic on the third plate – he put it at less than one hundreth of a milligram. Orfila threatened to sue Raspail for libel but none of this helped Marie. Sentenced to hard labour for life, she was eventually released on grounds of ill health after twelve years and died a few months later.

A strange incident occurred two years after the Lafarge case that attracted little publicity. At the trial of another woman accused of arsenic poisoning, Madame Lacoste, the analyst giving evidence, M. Chevallier, a colleague of Orfila, was asked whether the quantity of arsenic found in the Lacoste analysis was about the same as that which convicted Marie Lafarge. Chevallier replied: 'I cannot reply to a question so put; what was said to be the poison found in the body of Lafarge was imponderable: it was so infinitesimal that it could not fulfil the conditions of a standard of comparison when we use the words "more" or "less".'

L'Affaire Lafarge established Marsh's process as the seminal test for arsenic but at the same time it showed just how wrong he had been when he called the test simple and declared that any amateur could carry it out. So a year after Marie Lafarge's conviction, toxicologists were delighted to hear of a supplementary test that was said to be quicker and simpler. The German chemist Hugo Reinsch proposed boiling the suspect liquid with hydrochloric acid and a sliver of clean copper foil. Arsenic (as well as other metals) would form a grey-black metallic stain on the foil. 'Nothing can be more easy,' announced the toxicologist Robert Christison.

While the Reinsch test was simpler to perform than Marsh, however, it wasn't capable of detecting arsenic in all its forms and Marsh, with Berzelius's refinement, was better at detecting the quantity of poison involved: in a criminal case it clearly helped to know whether the arsenic present amounted to a lethal dose. And although Reinsch was an important advance in forensic science, it too was open to error, as a notorious court case was later to show.

Difficulties aside, in the view of the Society of Arts, Marsh had 'conferred an exceedingly great public benefit by his process' for 'in judicial examinations on suspicion of poisoning with arsenic, a chemist is usually called in for the purpose of examining whether the presence of arsenic can be detected in the contents of the stomach, the ejections from the intestines or in what may remain of the solid or liquid vehicle [usually food of some kind] in which the poison was supposed to have been conveyed', the Society said. If only a minute quantity of arsenic was present, however, then the processes involved required 'considerable familiarity with chemical manipulation', which meant the testimony of chemists was often 'uncertain and wavering'. This was to the injury of society if a guilty person were to escape justice and to the injury of the innocent 'if he goes away with a slur on his character from the ambiguity of a chemical test'.

The country at large shared the Society's relief that no longer would a guilty person escape justice or an innocent one be wrongly condemned. Some commentators went even farther: now that scientists could detect with absolute certainty the smallest amount of poison, no one would risk using arsenic as a means of murder again. The benefit that chemistry had conferred upon mankind could not be too highly estimated, the *Pharmaceutical Journal* declared. Now arsenic poisoning, this 'most execrable of crimes', was 'happily banished from the world'. Or so it seemed.

~ 16 ~

What if the Chymist
Should Be Mistaken?

The *Times* was not the only paper to denounce Sir
Edward Bulwer-Lytton's 1846 three-volume shocker
Lucretia, or the Children of the Night. The *Daily News*
called it 'hateful from the first page to the last', while the
Morning Chronicle accused the author of diving 'among
the dregs of crime' and being 'overpowered by the foetid
atmosphere'. The reviewers were particularly upset that Sir
Edward had presented his murderous heroine as educated,
intelligent, even sympathetic.

Given the subject matter and Bulwer-Lytton's lurid style,
the attacks in the popular press were predictable enough,
but the usually sober *London Medical Gazette*, edited by the
seriously scientific Alfred Swaine Taylor, sounded equally
overwrought, calling the novel 'a handbook on poisoning'.
Lucretia seemed to have been written almost for the express
purpose of 'giving a dignity to the crime of assassination
and of reviving in the public mind an interest in the lost art
of Italian poisoning'. The most eloquent novelist of the day
had given the world a work of fiction 'the entire plot details

and moral of which form a most complete revelation of the art of murder by poison'.

The accusation was somewhat wide of the mark. A budding murderer turning to *Lucretia* as an instruction manual would be disappointed. Sir Edward writes about a colourless, tasteless liquid that he calls the 'celebrated Neapolitan poison' (perhaps with Aqua Tofana in mind), which he claims had 'hitherto baffled every known and positive test in the posthumous examination of surgeons'. In the unlikely event that some 'unconjectured secret in the science of chemistry' were to detect it, no suspicion was likely to fall on the 'ministrant of death'.

No such poison was known in Britain, but by suggesting the possibility Sir Edward was encouraging the weak and the wicked to try their hand, the critics said. In fact, Bulwer-Lytton claimed that the crimes in *Lucretia* were based on real events: 'There has been no exaggeration as to their extent, no great departure from their details. In the more salient essentials ... I narrate a history, not invent fiction.'

Readers were quick to identify the real-life criminals. Marie Lafarge was deemed the model for the refined Lucretia Clavering, while Lucretia's stepson Gabriel Varney was said to be based on the British artist and writer Thomas Wainewright, who almost certainly poisoned his mother-in-law and sister-in-law for their life insurance, probably with strychnine. Wainewright was never charged with murder but in 1837 he was convicted of forgery and transported. His story haunted several writers, inspiring a slew of books and plays of differing merit. Perhaps, like Sir Edward, the authors saw glamour in his dashing persona. Charles Dickens spotted Wainewright on a visit to Newgate and used him for characters in *Martin Chuzzlewit* and *Little Dorrit*, as well as a short story, 'Hunted Down'.

The fact that the *London Medical Gazette* chose to discuss *Lucretia* at all was significant. The work hardly sat well with the texts the journal usually selected for review; recent titles had included *Clinical Observations on Diseases of the Genito Urinary Organs*, *Artificial Mineral Waters*, *Bathing and Sea Baths* and *On Gout: Its History, Its Causes and Its Cure*. What singled out *Lucretia* was the timing: its publication coincided with an outburst of public hysteria over poisoning that was to last for over twenty years.

For centuries the British had regarded poisoning as one of the black arts largely carried out by dastardly foreigners such as the Borgias and Giulia Tofana, but even before George Bodle died perceptions were beginning to change. Poisoning was no longer seen as rare and exotic but as a palpable threat to respectable citizens in their own homes. Back in 1827, in a strange, blackly humorous essay, 'On Murder Considered as One of the Fine Arts', Thomas de Quincey had lamented a retrograde step in the way the British killed each other: 'Fie on these dealers in poison, say I: can they not keep to the old honest way of cutting throats, without introducing such abominable innovations from Italy?'

The concerns were not completely unfounded. The number of suspected poisoning cases coming to court certainly went up at the height of the panic in the 1840s: between 1839 and 1848 the Old Bailey saw more than three times as many such trials as in the previous ten years – twenty-three compared with seven, while in 1809–18 there had been just two and in the twenty years from 1779 to 1798 none at all. And of the nineteen trials for arsenic poisoning at the Central Criminal Court between 1739 and 1878, nine occurred between 1752 and 1844, with eight more in the brief period from May 1846 to August 1850. Far from being frightened off by advances such as the Marsh and Reinsch tests, it seemed poisoners were becoming ever more ingenious.

Poisoning was a secretive, sinister business, and the ready availability of arsenic made it the most secretive, sinister substance of all. Whom to trust? Was that seemingly loving relative or loyal servant concealing a horrible purpose behind their smiles? Was every doctor as clever as John Butler in distinguishing a poisoning from a disease? And if poisoning were suspected, would the authorities be at all capable of bringing the criminal to book? No one had paid for the murder of George Bodle, after all. Most frightening of all was the thought that no one knew just how many people were getting away with it. If George Bodle's killer had been a little more subtle, a little more patient, would anyone ever have suspected that a frail old man with a stomach complaint nearing the end of his natural life had, in fact, been murdered?

'The science of poisoning seems to have now been brought to so high a point of perfection that its operation baffles the eyes of ordinary observers,' *The Times* believed. 'If you feel a deadly sensation within and grow gradually weaker, how do you know you are not poisoned?' asked *The Leader*. 'If your hands tingle, do you not fancy it is arsenic? ... Your friends and relations all smile kindly upon you; the meal ... looks correct but how can you possibly tell there is not arsenic in the curry?' Of course you can't, and the idea was terrifying. 'We dread it [arsenic poisoning] at nights ... we enquire actively as to its presence in our bread, and in our wine, and in our sauce,' claimed Henry Morley in *Household Words*.

In truth, Morley's chances of falling victim were slim. When criminal poisoning cases were at their height in the 1840s, there were still only ninety-eight such trials in England and Wales in a population of nearly twenty million people. And the threefold rise in poisoning trials at the Old Bailey in 1839–48 turned out to be a peak. In 1849–58 the figure went down to seventeen, before dropping back to seven in 1859–68.

Some of the increase in the 1840s was due to better detection. By enabling scientists to expose murders that might otherwise have gone undiscovered, James Marsh had unwittingly helped promote the idea that poisoners were everywhere and danger lurked in every sip of milk and mouthful of stew. But Marsh was not the only factor coming into play.

In 1836, the same year that Marsh published his paper, the government dropped the tax on newspapers from 4d to 1d. This, in turn, coincided with a rapid increase in literacy rates among the working classes, opening up a mass market for cheap, breezy reading matter. Stories of crime and punishment were as popular then as now, and poisoning cases were reported with particular gusto.

The upmarket papers also found that criminal poisoning was good for business. In 1856, when the *Illustrated Times* published a special edition on the trial of the 'Rugeley poisoner', Dr William Palmer – whose victims were said to include his wife and several of his children as well as a gambling friend, John Parsons Cook – circulation was said to have doubled to 400,000. And as Palmer awaited execution, hoping for a reprieve as the toxicologists argued over whether his alleged victims had died of antimony or strychnine or natural disease, *Lloyd's Weekly* ordered two 'gigantic' six-feeder rotary printing presses from Hoe of New York capable of churning out 15,000 copies an hour.

The first issue of the *News of the World* appeared on 1 October 1843. At 3d a copy, the paper aimed to be 'a novelty in newspaper literature' that the poorer classes of society could afford. On the front page, alongside an editorial promising 'the fearless advocacy of truth', was the headline 'Extraordinary Charge of Drugging and Violation'. The story concerned Emma Munton, 'a very handsome young woman of 17', who was said to have been sedated and abducted by a surgeon. The adjacent article was headed 'A Woman Poisoned by Endeavouring to Procure a Miscarriage'. William Haines, 'the proprietor of a respectable oil shop' (presumably there were some disreputable oil shops around), was accused of accidentally killing his wife with 2 ounces of potassium sulphate.

While the *News of the World*'s poorer classes of society, and many of the richer classes too, were enjoying a shudder over such cases, some of the more thoughtful or nervous citizens were worrying that the papers were becoming, like *Lucretia*, instruction manuals on murder. The budding criminal could study the mistakes that had put the accused in the dock – and avoid them. When the *News of the World* proprietor Lord Riddell ran into Frederick Greenwood, editor of the *Pall Mall Gazette* – a paper that described itself as being written 'by gentlemen for gentlemen' – Greenwood admitted that he had never seen Riddell's more rumbustious publication. Riddell sent him a copy. The next time they met Riddell asked Greenwood what he thought of it. 'I looked at it,' the editor said, 'and then I put it in the wastepaper basket. And then I thought, "If I leave it there the cook may read it", so I burnt it.' Perhaps it was significant that, of all his household, Greenwood was most worried about the woman who prepared his food getting her hands on the paper.

And every stratum of society was affected, or so it seemed. The Bodle and Lafarge cases concerned well-to-do, outwardly respectable people, but the burial club scandals featured those at the other end of the social scale. The 1830s saw the rise of the friendly society, a cheap version of the life insurance policies bought by the wealthy, and hand in hand with the friendly societies went a more specialist scheme, the burial club. In return for a subscription of about a halfpenny to a penny a week, the burial club would pay the insured person's funeral expenses. This way even those who had had a lifetime of indignity could avoid the ultimate shame of being buried in a pauper's mass grave. When an economic slump led to a massive rise in unemployment, however, a horrible practice was uncovered.

In 1843 the social reformer Edwin Chadwick reported that parents were enrolling their children in several burial clubs at a time in order to claim a large payout when they died. And some, he claimed, were going further: he had heard of cases where arsenic had been found in children's stomachs.

Benjamin Disraeli's novel *Sybil*, published in 1845, has a mother saying: 'This poor babe can't struggle on much longer. It belongs to two burial clubs: that will be three pounds from each, and after the drink and the funeral, there will be enough to pay all our debts and put us all square.'

But when Rebecca Smith was hanged at Devizes in 1849 for poisoning her one-month-old son Richard, no burial club money was involved, just relentless misery and hopelessness. Rebecca was forty-three, married to an alcoholic, said to be in poor health and 'suffering great deprivations'. She had had eleven children in eighteen years of marriage, but only the firstborn, a girl, had survived. When suspicions were aroused over Richard's death, the bodies of some of the other children were exhumed and William Herapath, the Bristol toxicologist, was called in to test the remains. One child, Sarah, had died in 1841 aged twenty days, and another, Edward, in 1844 aged fifteen days. Herapath found arsenic in their skeletons and in the black mould clinging to their bones. It was, he believed, the first time that arsenic had been discovered in a body that had been buried for eight years, and he brought his results – glass tubes of white arsenic, metallic deposits, Scheele's green and orpiment – into court so that the jury could see for themselves.

Rebecca meanwhile had confessed that she had indeed murdered her other seven children with arsenic. *The Era* printed a gratuitous detail, which fixed her story as a perversion of everything that motherhood was supposed to be. She had administered the poison, or so the paper claimed, 'in a manner the most unnatural' by applying arsenic to her breast, thus 'converting the channel of their [the children's] sustenance into the means of their destruction'. Rebecca's motive, she said, was fear that her children 'might come to want'. And her only worry about being hanged, she said, was that her husband would neglect her one surviving daughter.

Lloyd's Weekly reported her inevitable sentence under the

headline: 'Another female poisoner condemned to death', while the *Daily News* linked her story with that of Mary Ann Geering, convicted at Lewes for murdering her husband. So, in the perceived poisoning epidemic engulfing the country, it was women – from Eve with her tainted apple, through Augustus Caesar's wife Livia, Giulia Tofana and the Lucret(z)ias, Borgia and Clavering – who were most to be feared. They were scheming, duplicitous and, though physically and politically weak, exercised terrifying power through their domination of the kitchen and the sickroom. Like poison itself, women operated at a subliminal level, their purpose hidden until it was too late.

Fear of women poisoners had gone off the scale three years earlier when a perceived deadly sisterhood was uncovered in Essex. If Marie Lafarge personified the menace that lurked behind a sweet face and an elegant figure, then the likes of Sarah Chesham and Mary May pandered to another stereotype, the brutalised creature from the lower orders.

Sarah, dubbed Sally Arsenic, was first arrested on suspicion of poisoning a farmer's illegitimate baby for cash but soon the whispering began about the death of two of her own sons a few months earlier. The boys were exhumed and Alfred Swaine Taylor told the inquest that Joseph's body contained enough arsenic to kill an adult and James's enough to kill between three and six. Taylor produced their stomachs in court – Joseph's was intact but James's was in a 'soft, pulpy, cheese-like condition' – and he also showed the jury metallic arsenic on copper foil, presumably from the Reinsch test, that he had obtained.

The Times, while conceding the inconvenient fact that Sarah had yet to be found guilty of anything, branded the whole community complicit in her crimes: '... in an ignorant and secluded village of Essex lived a reputed poisoner – a woman whose employment was as well known as that of a nurse or washerwoman – who could put any expensive or disagreeable object out of the way ... an accepted and reputed murderess walked abroad ... unchallenged and unaccused' and 'all the inhabitants had seen her children buried without remark or outcry though they were clearly convinced that there had been foul play'.

When Sarah did stand trial for the deaths of her sons and the baby, to the surprise of many, not least presumably the editor of *The Times*, she was acquitted. Despite some lurid allegations – she was said to have skulked around the village with poisoned sweets in her pocket – the prosecution could find no evidence that Sarah had ever possessed arsenic, much less given any to Joseph and James, although someone had certainly done so if Professor Swaine Taylor was to be believed. But the story still had some way to go.

Three years later, Sarah's husband Richard sickened with vomiting and pains in his chest and abdomen. After his death, a post-mortem found some slight reddening of the intestines but his lungs were full of tubercules (nodules), partly ulcerated and sticking to the adjacent parts of the body. The chest cavity contained at least 3 pints of fluid. The doctors concluded that Richard had died of consumption.

Superintendent Clarke of the Essex Constabulary, who, said *The Times*, 'displayed much zeal and activity in the matter', asked Swaine Taylor to analyse some rice found in Sarah's kitchen cupboard. It contained no less than 12 to 16 grains of arsenic, Taylor said. He also found traces in Richard's stomach but not enough to account for death, although arsenic in small doses would prove more deadly in someone already weakened by tuberculosis. Mr Collins, the magistrates' clerk, asked Taylor about the chances of a conviction. Impossible on the medical evidence alone, said the toxicologist, although if the vomit had been analysed at the time it might have been a different story.

The case was held up as a perfect example of what happened when the finer points of poisoning were revealed in novels and newspapers. At Sarah's first trial a medical witness had described the effects of arsenic when administered under different conditions and circumstances, *The Times* reminded its readers. Sarah had 'stood quietly at the bar, listened and learnt'. And the *Pharmaceutical Journal* changed its mind. Now, far from rejoicing that arsenic was 'happily banished from the world', the publication began fretting about 'the modern assassin who adapts his expedients to the refinement of the age, bringing to his aid the appliances of science'.

If Sarah had indeed perfected her craft, it did her little good. Local feeling against her was now running high so the magistrates 'then resolved upon making an effort to solve the mystery'. A new witness came forward: Hannah Phillips claimed Sarah had offered to kill her abusive husband for her by 'seasoning' a pie. The Home Secretary, Sir George Grey, referred the matter to the attorney- and solicitor-general, who decided that Sarah had a case to answer. She was arrested, put on trial once again and this time Sarah was found guilty.

But in the interval between the deaths of Sarah's sons and that of her husband, there had been more strange goings-on among the women of Essex. On 14 August 1848, thirty-one-year-old Mary May, 'a repulsive-looking woman', was executed for the arsenic poisoning of her half-brother in a village called Wix, near Harwich. Her sufferings on the end of the rope were mercifully short owing to the fact that she was so huge and heavy, the papers reported. The motive was said to have been a small payout – £9 or £10 – on an insurance policy. A month later Hannah Southgate from Tendring in Essex stood trial for poisoning her first husband, Thomas Ham. Describing Hannah as 'an intimate friend' of Mary May, *The Era* said the two women were thought to be 'engaged in more than one of the poisoning cases which have disgraced this country'. Magistrates were said to be asking the Home Secretary to authorise exhumations in four other Essex villages.

Blacksmith Thomas Ham had lived a 'very miserable life' with his wife, who was accused of 'imprudent conduct' with a young farmer. After Mary May went to the gallows, the rumours about Hannah started in earnest. One woman reported hearing Mary say to Hannah: 'If he was my husband I would give him a pill' and Hannah replying: 'I'll be —— if I don't give him a dose one of these days'. When Thomas then died Mary was said to have visited Hannah to congratulate her.

Soon after, Hannah married her young farmer. Now Thomas was exhumed and once again Swaine Taylor was called in. This time he reported finding 15 grains of arsenic in the deceased's stomach, 'sufficient to destroy five full grown persons'. But unlike Sarah Chesham and Mary May, Hannah could afford a lawyer. William Ballantine, a practised Old Bailey hand, was retained. At Essex Assizes, Mr Ryland, for the Crown, began by admitting that there was not a scrap of direct evidence against the prisoner, but such crimes were not committed 'at times and places and under circumstances to enable witnesses afterwards correctly to give details', he explained. Poisoning was a crime which 'was kept secret, so that it was only under the circumstances, and expressions and acts that the crime could be proved'.

Many of the 'circumstances, and expressions and acts' that Mr Ryland was relying on consisted of local people describing rows between Hannah and Thomas and overhearing her threatening to murder him. One of the most damning witnesses was Hannah's one-time servant, Phoebe Head, who had a tale to tell about the state of the couple's marriage, her mistress preparing 'gruel or arrowroot or sago' for the victim and the dreadful sufferings he had endured. But Mr Ballantine exposed Phoebe as hardly impartial in the matter: Hannah had recently sacked her for stealing. When Phoebe was questioned about her three children by three different men 'she attributed her ruin to the prisoner, who engaged men to come to the house, but on being pressed, she admitted she had had a child before she went into Ham's house', said the *Ipswich Journal*. Whatever had being going on, Hannah had not wasted her savings on Mr Ballantine: she was acquitted.

The Essex cases were evidence of 'a moral epidemic more formidable than any plague which we are likely to see imported from the East', announced *The Times*, in a reference to Asiatic cholera. And there was some truth in the claim that poisoners were more likely to be women, even allowing for any gender bias in reporting. Between 1820 and 1839, fewer than half (44 per cent) of *The Times'* accounts of poisoning trials featured women as the accused, with virtually equal numbers of men and women accused of trying to poison their spouses. During the 1840s, however, nearly two-thirds (60 per cent) of the stories involved women defendants.

The day before Sarah Chesham was hanged – 24 March 1851 – the Earl of Carlisle moved the third reading of his Sale of Arsenic Regulation Bill in the House of Lords. In view of the vast number of suggestions he had received, he was proposing several amendments. Mostly they involved unexceptional tweaks to the proposed regulations. Now, as well as the druggist keeping a record of the purchaser's name, address and reason for buying arsenic, for example, the buyer would have to sign what became known as the poison book. But then in the middle of a tedious discussion about plans to colour the substance, the earl slipped in a quick mention of a more fundamental proposal. He 'thought it expedient that it should be expressly enacted that arsenic should be sold to none but male adults'. So, like children, women were to be banned from buying arsenic. Their lordships approved the amendments and the Bill was duly sent to the House of Commons for the next stage of its passage into law.

If Carlisle hoped his sleight of hand would go unnoticed, he would be disappointed. His machinations came to the notice of the philosopher and fearsome campaigner for the rights of the individual, John Stuart Mill, who wrote a strident letter to the Home Secretary. The clause amounted to all women, from the highest to the lowest, being declared unfit to have poison in their possession, he claimed. It was impossible to believe that 'so monstrous a proposition' could have received government approval except through 'inadvertence'.

'On what supposition were men to be trusted with poisons and women not, unless that of their peculiar wickedness?' Stuart Mill asked. 'And for what reason, or under what incitement, is this insult passed upon them? Because among the last dozen murders there were two or three cases ... of poisoning by women?' If the last two or three murderers had been men with red hair, Parliament might just as well have rushed to pass an Act restricting all red-haired men from buying or possessing deadly weapons, he declared. On 5 June the Sale of Arsenic Regulation Bill received Royal Assent. But the 'men only' clause had been removed.

Alfred Swaine Taylor had played a central role in the Essex cases. By the 1840s he and Henry Letheby in London, Robert Christison in Edinburgh, and William Herapath in Bristol were making regular appearances in the witness box. Their presence in the courtroom began to signify the importance of the case. They arrived with their specimen jars, glass tubes and porcelain dishes to show the jury the tangible evidence of their findings. Unlike the local surgeons and apothecaries whom the courts had relied on in the past, these men were specialists in their field. Unfazed by legal procedure or the solemnity of the occasion, they were more than capable of holding their own against tricky lawyers. The public could rest assured: here was a new breed of scientists, possessed of cutting-edge knowledge that enabled them to prove a murderer not only guilty beyond reasonable doubt, but beyond any doubt at all.

William Herapath had gained a national reputation when he gave evidence at the trial of Mary Ann Burdock, accused of poisoning her lodger, Clara Smith. Clara had been buried for fourteen months but Herapath's three precipitate tests clearly showed the presence of arsenic in her body, and Mary Ann was convicted. The exhumation was said to be the first in England to be carried out for the purposes of chemical analysis, and such was the interest that the *Bristol Mercury* rushed out a special issue carrying the evidence verbatim so that readers could 'forward it to their friends in distant parts'.

Like Herapath, Henry Letheby became best known through a high-profile murder case. Ann Merritt was accused of killing her husband James, a turncock with the East London Water Company, with arsenic. James had been in a burial club and early on the morning he died, his wife had claimed £3 15s in part-payment of the total £7 10s sum she had coming. Had James lived for another ten days, though, the payout would have gone up to £10, which mitigated in her favour. Ann said that James must have taken the arsenic in mistake for indigestion powders. She had bought the poison herself in order to commit suicide, she claimed. (A neighbour ruled out the possibility of James having committed suicide on the grounds that he had just bought new boots.)

The inquest jury decided that James had died from 'the deleterious effects of a mortal poison, called white arsenic, found in his stomach; but when, how or by whom, or in what manner the same has been administered and taken ... there is no satisfactory evidence to prove'. Despite the inquest verdict, Ann was charged with murder and Letheby testified at her Old Bailey trial to finding 8½ grains of white arsenic in James's stomach contents, 'much more than sufficient to have caused death', as well as one fourteenth of a grain in his liver and a trace in his intestines. His stomach contents resembled thick gruel, Letheby said, adding that some of the arsenic was administered not more than two or three hours before he died. This was damning because by then James had been too ill to have taken the poison himself, by accident or design, and because Ann was the only person with him at the time.

The jury duly found her guilty, but as Ann awaited execution in Newgate prison, a London surgeon called Davies wrote to the *Daily News* to take issue with Letheby that the liquid in James's stomach had been ingested only two to three hours before death. Some perfectly healthy people had slow digestions, he said. Had Letheby analysed James's stomach contents to prove that they contained the remnants of food, as he claimed? If not, then perhaps Letheby had mistaken the supposed 'alimentary matter' for a mixture of gastric juices and normal mucous membrane that had been softened by the poison.

He was not the only person feeling uneasy. Another *Daily News* reader, signing himself 'Alpha', criticised Letheby's claim about digestive transit times and cited a case in 1847 where he said Letheby had wrongly claimed that arsenic burned with a blue flame. Had no one come forward to challenge that assertion, an innocent man would have hanged. In fact, elemental arsenic does burn with a blue flame but white arsenic does not. It is not clear what form Letheby was dealing with. 'Alpha' was quickly followed by 'Beta' and 'Gamma' and other correspondents, all lining up to accuse Letheby of trying to hurry Ann Merritt to the scaffold on the basis of dodgy science.

Rattled, the authorities stayed the execution while three eminent medical men reviewed the evidence. They all disagreed with Letheby, who hit back furiously, saying that because of their 'total inexperience of the action of this particular drug, they were not among the best qualified to given an opinion upon the toxic effects of arsenic'.

The Liberal MP John Bright raised the case in the House of Commons and Letheby then turned on him, accusing Bright of using Merritt to advance his campaign for the abolition of the death penalty. But in a letter to a medical journal, after taking up two and a half columns of newsprint justifying himself, Letheby finally conceded that it was 'just possible' for the sequence of events 'to have happened otherwise [than he had stated in court]' and it was 'even possible' that the poison 'may have been taken as the condemned woman asserts'. Ann's death sentence was lifted: she was sentenced to transportation instead for her carelessness in leaving arsenic in an unmarked packet lying around next to the stomach powders.

Few people were arguing that Ann Merritt was definitely innocent, rather that she should not have been condemned on the opinion of one man, however experienced in his field. The case conjured up a different spectre to that of the ingenious poisoner running rings around the scientists. Was it possible that those scientists were sending innocent men and women to the gallows? It would appear that the results of the latest chemical analyses were not, after all, objective and indisputable matters of fact but shifting and insubstantial, sometimes depending on little more than a reading of the runes by a human being who, for all his impressive words and the paraphernalia that he brought into court, might still be wrong. Had anything much changed from the days when analysts threw arsenic on the fire and sniffed for garlic?

By the time Thomas Smethurst stood trial chemists had

had over twenty years' experience with the Marsh test and seventeen years with Reinsch, but that was still not enough to prevent a debacle. Smethurst, a forty-eight-year-old doctor with a hydrotherapy practice – 'a man of small and insignificant mien with a reddish-brown moustache' – left his elderly wife and went through a fake marriage with Isabella Bankes, who was in her forties and had a private income and a large life insurance policy.

The pair moved into lodgings on Richmond Green in Surrey, and a few months later Isabella began to suffer bouts of vomiting and diarrhoea. Despite being seen by three doctors, she failed to respond to treatment and the medical men began to wonder whether something other than a natural disease was going on. They sent a sample of her vomit to Alfred Swaine Taylor, who said his tests recovered a metallic deposit that he thought was either arsenic or antimony. Smethurst was promptly arrested. Isabella died the next day.

Tests on the post-mortem samples revealed no arsenic but did find a trace of antimony in the intestines and one kidney. Taylor and a colleague from Guy's Hospital, William Odling, then turned their attention to the large number of medicine bottles in Isabella's bedroom, and here they ran into a little trouble. All the potions appeared innocent except one. When they ran the Reinsch test on the contents of bottle number 21, they found to their surprise that the copper gauze used to collect any traces of arsenic dissolved when they added it to the solution. They dealt with the problem by continuing to add gauze after gauze until finally the liquid could dissolve no more. They then introduced a new copper gauze which 'at once received the arsenic'.

Bottle 21 contained chlorate of potash, an innocent

diuretic, mixed with arsenic, Taylor told the Richmond magistrates. He had never before come across such a combination and there was no medical reason for prescribing it. Smethurst was duly charged with murder. Leading the prosecution team was William Ballantine, who had saved Hannah Southgate's neck eleven years earlier, and assisting him was William Bodkin, who had helped defend Young John Bodle at Maidstone Assizes.

Seven doctors appeared for the defence and ten for the prosecution, among them Alfred Swaine Taylor. On the second day of the trial, however, Taylor was forced to enter the Old Bailey witness box with an embarrassing tale to tell. He had carried out more experiments and consulted colleagues and a consensus had been reached: 'From the copper I used in my experiment on bottle number 21, I actually deposited myself arsenic in the liquid that remained,' he told the court. Arsenic was known to be an impurity in raw copper but scientists believed it was eliminated during the refinement process. Clearly this was not always so and the chlorine in the chlorate of potassium had dissolved the copper, thereby freeing the arsenic. Still, Taylor told the court, from the description of Isabella's symptoms and from the post-mortem appearance, he could ascribe the death 'to nothing but the action of irritant poison'.

Summing up for the defence, Mr Parry referred to 'the terrible mistake made by Dr Taylor'. This was the man who had stated in the magistrates' court 'distinctly and positively and without reservation' that he had found arsenic in a bottle under Smethurst's control. That fact alone would have been enough to send the accused to the scaffold.

Even so, the jury clearly didn't like the cut of Smethurst's jib and found him guilty. Certainly he had behaved suspiciously, refusing to let Isabella's sister see her during her last few days and dragging a solicitor out on a Sunday as she lay dying in order to draw up her will, which stated: 'all my real and personal property, estate and effects whatsoever and wheresoever and of what nature or kindsoever the same may be, I give, devise and bequeath the same unto my sincere and beloved friend Thomas Smethurst'.

A huge outcry followed the verdict, however. The general consensus was that Smethurst, though obviously a bounder, was not a proven murderer. Thirty doctors wrote to the Home Secretary to say 'there is no proof worthy of belief that arsenic was found in the evacuations or remains of the deceased and still less that death was produced by poison', while three of the medical men who appeared for the defence wrote that Isabella's symptoms and pathology were not consistent with arsenic or antimony poisoning, but were consistent with dysentery in a pregnant and previously unhealthy woman and that her death was 'fairly ascribable' to such a cause. (The post-mortem had revealed a foetus of between five and seven weeks' gestation.) Twenty-eight barristers also protested that the evidence did not warrant the verdict.

The Home Secretary asked Sir Benjamin Brodie to review the case. 'Although the facts are full of suspicion against Smethurst, there is not absolute and complete evidence of his guilt,' Brodie concluded. As for the toxicology, there was no proof of Miss Bankes having taken arsenic 'and indeed many of the symptoms … arsenic produces were absent'. Thomas Smethurst received a reprieve four days before he was due to hang and was charged with bigamy instead. Explaining that he would be recommending a free pardon for Smethurst as well as the reprieve from execution, the Home Secretary said the necessity arose 'from the imperfection of medical science and the fallibility of judgement … even of skilful and experienced medical practitioners'. The heavens then fell upon the hapless Swaine Taylor.

The *British Medical Journal* pointed out that, had Smethurst had been tried on Taylor's initial evidence, his life would have 'fallen through this illusive gauze wire'. The *Dublin Medical Press* said that the man without whose assistance no one in England could be found guilty of criminal poisoning had brought his profession into a disrepute that years would not remove. All that remained for him now was to withdraw into obscurity, not forgetting to take his arsenical copper with him.

The mud-slinging within the profession was on an industrial scale. Herapath professed himself astounded at Taylor's casual comment that he had used the same piece of copper in the Reinsch test for twenty years. 'What shall be said of the justice of the convictions and executions which have taken place during those years upon Dr Taylor's evidence?' he asked. And he went on to criticise Swaine Taylor's handling of the entire analysis: 'The whole set of operations were a bungle. Reinsch's process is not applicable where nitrates or chlorates are present.' Taylor then found fault with some of Herapath's argument about arsenic testing and Herapath hit back with 'the doctor seems to fancy that he is competent to correct my errors'. His first duty was to correct his own, and then people might have some confidence in his ability to instruct others.

William Odling, whose fingers were literally and metaphorically all over the infamous bottle number 21 along with Taylor's, weighed in next. He was furious at Taylor's account of what happened. Describing the blunder as wholly indefensible, Odling said he was heartily sorry for his own part in it but that part was 'much less than what the public have been led to suppose'. He then explained at some length how the whole debacle was really Taylor's fault, while the credit for discovering the mistake was not in any way due to Dr Taylor but due 'solely to me ... when experimenting in my own laboratory'.

The editor of the *British Medical Journal* wished a plague on all their houses. 'Dr Taylor ... never allows an opportunity to escape of sneering at the scientific achievements of Mr Herapath and the latter ... gentleman is not behind-hand in returning the compliment. Dr Letheby is equally complimentary to Dr Taylor,' he wrote. The public watched 'as professional men stabbed each other's reputations over the bodies of malefactors'. Was there perhaps something in poison-hunting that bred such unnatural storms?

The Smethurst case set back the cause of forensic toxicology – for which so many over-confident claims had been made, some by those who should have known better – and damaged the credibility of the expert scientific witness for years to come. 'We must needs pin our faith on the conclusions of chymists and hang a fellow creature because a small crystal, so minute that it can only be recognised by the microscope, is exhibited on a scrap of copper wire,' *The Times* remarked, asking: 'What if the chymist should be mistaken? What if science is at fault ...? How are 12 men notoriously deficient in specialist knowledge to decide when professional witnesses disagree?' The question has never been satisfactorily answered.

17

The Freezing Influence of Official Neglect

James Marsh's Bodle-inspired discovery was the first major advance in modern chemical toxicology and his test remained in use, with some modifications, for nearly 150 years. The achievement did not prove a life-changing event for the man responsible, however. For although James Marsh acquired an international reputation after what became known as L'Affaire Lafarge it certainly did not make his fortune, and he continued to get by on his surgeryman's salary topped up by his fee for assisting Michael Faraday, some consultancy work for the Woolwich gas company and the odd award or gratuity. He and Mary remained in the small house in Beresford Street for the rest of their days.

In 1837, the year after he received his Gold Medal, the Society of Arts honoured Marsh once again, this time with a second silver medal and a prize of 30 guineas for the quill percussion tube for firing cannon he had invented five years earlier. The tube had immediately been adopted throughout the British Navy, with the army following nine years later. The Ordnance Board gave Marsh a bonus of £30.

Then, in 1839, HMS *Excellent*, the Royal Navy ship that had tested the percussion tube, tried out another of the surgeryman's ideas: a fuse that caused shells to explode the instant they hit their target. In this case, *Excellent*'s captain, Sir Thomas Hastings, concluded that the device had potential but was not yet reliable enough for use, and he recommended more experiments be carried out. That same year the Society of Arts gave James Marsh a final accolade: a vote of thanks, but no medal or money, for concocting a liquid made from the petals of the red dahlia as a test for acids and alkalines. This had an advantage over the then usual method, according to Marsh: 'The infusion of the common red cabbage has long been in use ... and, although possessed of great delicacy in this respect, is still subject to an objection on account of its becoming so exceedingly offensive in its smell after having been prepared a few months.'

And in 1840 he earned an extra £25 for giving chemistry lessons to a young smithy in the Royal Laboratory forge. The Ordnance Board proposed that 'Mr Marsh, a very clever chemist at Woolwich, should instruct a young man by name Tozer in the knowledge of chemistry as far as may be required for laboratory purposes'. The Board made sure it got its money's worth, for the firemaster Colonel Dansey also sat in on the lectures. This was the same man whose kite had won a Society of Arts gold medal when Marsh took the silver for his electromagnetic apparatus.

On 21 June 1846, James Marsh died unexpectedly at the age of fifty-two after a two-week bout of what his doctor recorded as inflammation of the liver (hepatitis can be acute, for example due to an infection, or chronic through alcohol damage). James Achindachy was at his bedside in

Beresford Street. When Achindachy registered his friend's death he gave his occupation as chemist, although in his will Marsh describes himself by his correct title as dispenser of medicines.

Marsh's will is an altogether simpler document than George Bodle's. The chemist, like the farmer, leaves all of his 'household goods, furniture, plate, linen, china, books, prints, pictures' to his widow, although Mary Marsh, unlike Ann Bodle, is described as his 'dear wife'. And unlike in George's last will and testament, there is no complex share-out of land, houses, outbuildings and bank securities; no messeges, tenements or appurtenances to worry about. Marsh had none to leave. He did have one possession but it was of little practical use to Mary, now deprived of a breadwinner. He left her 'all my chemical apparatus of every description used by me in the Arsenal or elsewhere'.

Mary was also to receive 'any monies, fees, gifts, dues, rewards for services, inventions or otherwise' but in reality this turned out to be about as much use as the laboratory equipment: either there were no such fees, rewards or dues for inventions owing when James Marsh died or the amount was so little it provided scant comfort for Mary Marsh. There was just one bequest to someone other than his wife and daughters, a sentimental gesture common at the time but one that George Bodle had had no truck with: 'I hereby give unto my friend James Achindachy a mourning ring as a testimonial of my esteem.'

James's death left Mary penniless and she had no right to a pension from her husband's employer. As James had never held the official post of ordnance chemist, he was not on the civil establishment but was paid by the day like a casual labourer. A few days after the funeral, then, Mary wrote to the Board asking for a discretionary pension in view of her husband's contribution to the Arsenal over the years, particularly his work on the fuse and the percussion tube. She had a few allies. Rear Admiral Thomas Dundas wrote to the Board reminding them that he had been the Clerk of the Ordnance when Marsh was first brought into the service. 'I think his widow well deserving a liberal pension, as Mr Marsh was a zealous officer and very badly compensated for his guns, etc.,' he said.

And after praising Marsh's expertise and 'valuable assistance', which 'tended greatly to the benefit of the public service', James Cockburn, director of the Royal Laboratory where Marsh had begun his career, wrote: 'From his very limited income it was out of his power to provide for his family. I therefore respectfully trust that the Hon. Board will take into their most favourable consideration the prayer of Mrs Mary Marsh, widow.' Three days later, Mr Ryham, secretary to the Board, received another letter from a group of Woolwich inhabitants: 'We ... consider Mrs Marsh's appeal to the Government for a pension in consideration of the importance of the inventions of the late Mr Marsh ... to be as reasonable in its nature as the necessity is urgent, and we therefore most respectfully and earnestly request you will be pleased to recommend her case to the favourable consideration of the Honourable Board of Ordnance.' The signatories included William Nokes, the Plumstead parish lawyer who had put the case against Young John at Coroner Carttar's inquest.

A helpful Board official decided to look back through the records to see whether there was any precedent for paying discretionary pensions. At first glance the situation looked promising: 'There are instances of widows being granted pensions who were not entitled under the regulations and those whose cases are so far analogous in as much as their husbands were not on the civil establishment of the Ordnance: viz. Mrs Pett, widow of Pett, master of the Ebenezer sloop, a pension of £20 p.a.; viz. Mrs Turn, whose husband was mate of The Earl of Chatham, a pension of £15, and also Mrs Mantle, widow of W. Mantle, master of The Duke of Wellington, a pension of £20 p.a.' But someone had then scribbled a note on the file: 'These pensions were granted in consequence of the husbands of the parties having been drowned at sea in the execution of their duty.'

In September the Board came to a decision: 'Mrs Marsh may be granted a donation of £20. The Board do not consider that she has any claim ... for a pension.' One month later, on a small piece of stationery edged with black, Mary replied: 'I hope Sir you will pardon me for stating that I do think the government has not done me justice.' The letter was accompanied by a formal petition explaining how she had been 'left by the sudden bereavement of Mr Marsh totally destitute of the means for the future support of herself and two children'. The money and medal Marsh had received for his percussion tube, although highly complimentary, were 'by no means an adequate remuneration for an invention that has been proved to be of great importance to the service'.

In the meantime, Marsh's elder daughter Lavinia wrote to the Prime Minister on her mother's behalf. She received this reply: 'Madam, I am requested by Lord John Russell to acknowledge the receipt of your letter of the 17th instant soliciting a pension for your mother as the widow of a scientific man. In reply I have to express to you his regret that he cannot hold out the prospect to you of his being able to forward your views as the want of funds would prevent him, even if he were able consistently with the stronger claims of others, to recommend Mrs Marsh to Her Majesty for a pension on the civil list.'

By now, though, the matter had come to the attention of the press. *The Times* was astonished: 'Mr James Marsh, the celebrated chemist whose well-known test for the detection of arsenic is so extensively used in medical jurisprudence, died some short time since leaving a widow and family in very needy circumstances. On his death his widow memorialised the Board of Ordnance for a pension. The board have just sent her the donation of £20!'

And someone signing himself L.T. wrote to the paper from Brussels. He was not prompted, he said, by the 'many surprising attainments of Mr Marsh', nor was he writing in order to emphasise 'some of those innumerable contrivances by which he ingeniously combated the freezing influence of official neglect and stocked his humble laboratory', nor, even, to denounce 'the propriety of that miserable economy which has induced our government to award the paltry sum of £20 to the widow and family of such a man'. Instead, the correspondent merely wished to state that he had two guineas very much at the service of any gentleman who would undertake the collection of a subscription for Mrs Marsh and family. Only absence from England prevented him from doing so himself.

Moved perhaps by Mary's letters or, more likely, by the realisation that its treatment of her was making uncomfortable reading in the press, the Board of Ordnance began to have second thoughts. A note dated 28 October records: 'The Board submit to the Master General that having reconsidered the case of Mrs Marsh, they considered that ... she is entitled to more than a mere donation of £20. I therefore submit that a pension of £20 p.a. may be granted to her as a special case.' On 2 November, then, thirteen years to the day since George Bodle drank the cup of coffee that had killed him, the Master General appeared to settle matters: 'I concur with the Board. Let this case be recommended to the Treasury.' Mary had to wait another five months while the Board of Ordnance and the Treasury worked out which budget the money was to come from but she finally received her pension.

The niggardliness of the amount provoked more indignation. *The Examiner* denounced how 'men of genius' were rewarded for their public service, calling it 'petty larceny', and referring to 'those precarious and scanty doles which men of letters, in common with the menials of royalty, now obtain at rare intervals from the caprice of a Minister'. It was an outrage that the widow of a man who had an international reputation through his discoveries in chemistry had, after repeated applications, received a pension of only £20, *The Examiner* believed. If the reputation ascribed to Mr Marsh was deserved, then the 'paltry pittance' was 'ludicrously and ignominiously inadequate'.

This time the government stood firm, but fortunately L.T.'s appeal prompted donations from scientists and members of the public that eventually amounted to some £500. One contribution of five guineas came from Professor Michael Faraday, James Marsh's boss at the Royal Military

Academy, who had assigned him the Bodle analyses and in so doing sparked the discovery of what was to remain the gold standard for arsenic testing until well into the second half of the twentieth century.

~ 18 ~

I Went with a Lie in my Mouth

On 30 December 1843, ten years after George Bodle's death, a Woolwich auctioneer called Austin placed an advertisement in the *Kentish Independent* announcing the sale of a valuable freehold estate in Plumstead. The sale, in several lots, was to take place in January if the property had not been disposed of privately by then. It consisted of around fifteen acres of orchards, plantations and gardens, two good brick-built houses in the centre of the village and 3 acres of marshland. Details and plans would soon be ready for inspection at Mr Austin's place of business at the Western Life Insurance office or at the offices of the lawyer James Colquhoun of Rectory Place.

The property was not disposed of privately, for on 27 January Mr Austin was advertising again, this time announcing that the sale would take place at the auction rooms in Woolwich on 7 February, and adding that the land included ten plots of ground fronting the high street that were suitable for building, along with a large barn and brick stabling.

The estate belonged to the late John Bodle – that is, Middle John, who had been holding it in trust for his children George, Young John and Mary Andrews under the terms of his father's will. Middle John had stayed on in Plumstead for some three years after the trial. His step-mother Ann had died in July 1836 aged seventy-seven, at which point the rents on Middle John's children's share of old George's property passed from her to him, while the Baxters took immediate possession of their lands and investments. William Baxter in particular was a man of substance at the age of twenty-six.

At some point after April 1836 Middle John left Plumstead and by 1841 he was living 20 miles away in Northfleet, near Gravesend, with his daughter Mary, her husband Thomas Andrews, and their three boys. Despite being distraught during her brother's trial, Mary seems to have borne her father no ill will for his part in the proceedings. The Andrews family had clearly prospered, for the 1841 census describes Thomas no longer as a coffee shop owner but as being 'of independent means'. The couple also employed two live-in servants and owned at least two houses back in Clerkenwell which were rented out to tenants. (In 1842, a Gravesend builder called Thomas Carter was sent to prison for four months for stealing 24 pounds of lead from number 7 Clerkenwell Close, the property of one Thomas Andrews. Andrews had employed Carter to unblock a smoky chimney and paint the house but Carter had also carried out some unauthorised work, climbing on to the roof, stripping off the lead and replacing it with zinc.)

Middle John died aged fifty-eight on 17 October 1843 of what was then known as dropsy, a build-up of fluid in the soft tissues that is often associated with congestive heart failure. By then the Andrews had also acquired a Northfleet pub called the Windmill Tavern, and it was here that Middle John died. His last words, if a local journalist is to be believed, were the surprisingly eloquent announcement: 'I have been a profligate but I leave the world unstained by crime.'

On 5 February 1844, less than a month after the sale of the Bodle property, a thirty-three-year-old man named James Smyth appeared in the dock at the Old Bailey. He was charged with extortion and was said to have 'feloniously sent a letter to Thomas Robinson demanding money with menaces, without any reasonable or probable cause'. Six other counts relating to the same offence were on the indictment. The prosecuting counsel was one Thomas Clarkson, who had defended Young John Bodle to such good effect at Maidstone Assizes eleven years earlier. The forty-eight-year-old Robinson went into the witness box to tell his story. He was butler to Lord Abingdon and had first met the defendant by chance in the street in Brighton, where he was living with his master some three months earlier. They had fallen into conversation and after that had come across one another several times, stopping to talk. 'He did not know I was in service, I did not know how he was getting his living,' Robinson said. 'I never visited him or he me, we merely spoke casually in passing.'

On 2 January Lord Abingdon came to London to have an operation and was accompanied by some of his household. The party checked into the elegant Grillion's in Albermarle Street, Mayfair. The Frenchman Alexander Grillion had opened the hotel back in 1803 and it was here that Louis XVIII had stayed in some splendour before returning to France after the defeat of Napoleon. Along the street were the imposing Doric colonnades of the Royal Institution and at number 50 the offices of John Murray, the publisher of Byron, Jane Austen and, later, Oscar Wilde.

When the butler had been in the capital for about a fortnight, he was pleasantly surprised to run into his Brighton acquaintance in the Strand. 'He recognised me and said, "What, you have come from Brighton?" or something similar to that,' Robinson said. 'We entered into conversation … It was raining very hard. I asked if he would like a glass of ale, he said he should prefer a gin and water and I gave him one at a public house close to Charing Cross hospital.'

During this conversation, Smyth asked Robinson whether he would in due course take a parcel back to Brighton for him; it was a present for someone he had lodged with and who had been kind to him. Smyth gave him his card, showing the address of Mrs Walters' and Mrs Hunt's private lodging house, 116 Drummond Street, 'opposite the arrival side of Euston Square station'. He then asked where the butler was staying and the pair parted company. At about nine o'clock on the evening of Saturday, 27 January, Smyth turned up unannounced at Grillion's. 'There is no room or place of reception for the servants, except the front hall where persons are passing to and fro,' Robinson explained. 'I had a bedroom there … and he was shown up to my room. He did not bring any parcel for me, he said he would send it.' The pair chatted for about half an hour, Smyth talking mainly about his visits to Ireland and Liverpool. Robinson then showed him out.

The following Monday morning at about eleven o'clock, a waiter handed the butler a letter. It was from James Smyth, dated the previous day. It gave Smyth 'the greatest pain' to have to write, he assured Robinson, but he thought it better to communicate with the butler first before writing to his lordship, Robinson's master. During the course of his visit on the Saturday night he had lost a small brown pocketbook containing three £5 notes, either in Robinson's bedroom or on the stairs. Given that he had 'a call for a large demand of cash on Wednesday morning', the loss of the money would, he said, 'greatly injure me'.

But he suggested a solution. The butler might wish to replace the missing £15 himself without further fuss, for 'that dreadful insult offered by you on that night, in wishing me to let you commit such a crime on my person as you did, is much against you, and unless you forward to me the sum lost by Tuesday night, my solicitor will write to his Lordship on Wednesday morning concerning your base conduct'. He was, he told Robinson, 'bent on going to the utmost extremity of the law, should you fail in sending it to me at my lodging by the time named', adding: 'I feel so disgusted with you that I sincerely hope you will not attempt calling on me … Return me what I have lost and keep out of my way, and no one shall ever hear of it.'

'I never took or attempted to take, any liberty of any kind whatever,' the butler told the jury. 'I went downstairs with him and let him out. I should say he did not drop in my room any brown pocket book with three £5 Bank of England notes in it – he could not have done so without my finding it – he did not drop any on the stairs, nor were any such afterwards found.' It was a very shaken Mr Robinson who, ten minutes after reading the letter, sought out the hotel proprietor, M. Grillion, to ask his advice. The Frenchman was in no doubt about what to do.

On the morning of Wednesday, 31 January, the day after the deadline ran out, the butler received a second letter from Mr Smyth. The price had gone up. 'Unless the sum of £30 is sent to me by seven tonight my solicitor will be with his Lordship tomorrow morning,' the note ran. 'My word is my bond, such you will find. They were three 10 pound notes, not fives, and such I shall expect you to forward without delay. I give you my word that I will go to the very extent of the law. I cannot see you or anyone you may send nor shall I write again. I wish you to remember that you have not a low person to deal with but a highly respectable gentleman and one who can condemn you before the eyes of the world to be guilty of the most horable [sic] crime and will do it should you fail in returning to him the amount lost this night. If you enclose it in a small parcel it will come to my hand quite safe.'

That evening Mr Wright of Messrs Wright and Smith, attorneys at law of Golden Square, Soho, gave his clerk William Rabbeth some letters the firm had received from Grillion's Hotel, along with instructions. Rabbeth immediately set off for Grillion's for a talk with Mr Robinson and the pair then left the hotel and made their way to Vine Street police station, where the butler told his story. Their next port of call was in a rather more modest part of town – Mrs Walters' and Mrs Hunt's private lodging house at 116 Drummond Street, near Euston railway station. They were now accompanied by police officer C6, otherwise known as Sergeant Joseph Mount.

The Metropolitan Police archives fail to record a Joseph Mount but from December 1840 to February 1851 he crops up regularly in the Old Bailey trial records, giving evidence in humdrum cases of petty larceny and pickpocketing around Oxford Street, Piccadilly and Westminster. He had already come up against the lawyer Thomas Clarkson and had not enjoyed the encounter. Mount was appearing as a witness against a compositor called Richard Lobb, who had been arrested for stealing stockings, a scent bottle and other bits and pieces from a woollen-drapers in Leicester Square during a fire at the premises. Lobb knew the draper and claimed he was helping to rescue his friend's possessions. Clarkson cross-examined the policeman mercilessly, forcing him to contradict himself so many times that the judge commented: 'I cannot believe the evidence of that policeman,' and the case collapsed. Even so, by October 1843, Mount had been promoted to sergeant.

'I knocked at the door, and asked whether a person named Smyth lived there,' Rabbeth told the jury. 'I found that he did and was at home. I was asked into the passage and the prisoner came down to me.' Rabbeth introduced himself as a friend of Mr Robinson's and Smyth showed him into the back parlour. 'I told him I understood he had made a very abominable charge against Mr. Robinson,' Rabbeth went on. 'He said, "Hush, hush, there are gentlemen in the other room, they will hear what you are saying". I told him that rather than avoid any exposure in such a disgraceful matter, I had called to endeavour to arrange it.'

Rabbeth asked Smyth whether he had written the letter of 28 January, asking for £15. 'He said he had ... I then asked if he had written a letter the following day, in which he claimed £30. He said, "Yes, I certainly did". I asked why. He said Mr. Robinson had not acted in a gentlemanly manner, and that was the reason he had doubled his demand ... I said I thought it was a great pity a thing of this sort should become publicly known.'

Rabbeth then told Smyth that Robinson had only £5 to his name; he had tried to borrow the rest from a friend but failed. 'He then asked, if I was a friend of Mr. Robinson's, why I did not pay the money myself ... I drew my purse from my pocket and said, "All the money I have is about £7 or £8". I then produced a £5 note and said, "Will you take this on account and I will pay the other £25 in the course of tomorrow?" He said, "I have no objection to taking the £5, providing you promise me faithfully to bring the £25 tomorrow". I asked what time it would be convenient ... and one o'clock was fixed. I gave him a £5 note of the Brighton Union Bank.' As this was going on Sergeant Mount was in the street watching the house from the other side of the road while Robinson hovered in a doorway a couple of houses down. The business concluded, Rabbeth asked Smyth to show him out. 'He came to light me to the door. He put the light down on the ground. I took off my hat, which was the signal agreed upon with the policeman,' said the clerk, adding: 'I went with a lie in my mouth in order to catch him.'

Mount promptly ran across the road and collared Smyth, while Robinson emerged from hiding and identified him. The prisoner broke out of the policeman's grasp, however, and made a dash for it, running back into the hotel with Mount and Rabbeth in hot pursuit. The two men chased him into the parlour, where he appeared to throw something away. 'It was the £5 note I had given him,' Rabbeth said. 'He said, "Hush, hush, don't make a noise, this thing may be settled".' The thing was indeed settled but not in the way that Smyth had hoped: he was taken to Vine Street police station and charged with extortion.

On 1 February Smyth appeared at Marlborough Street Police Court, where his case was one of the more serious to come before the magistrate, John Hardwick. His fellow accused that day included William Adams, described as a 'lame young man', accused of wilfully damaging a picture in the National Gallery. A witness described how Adams had struck Mola's *Jupiter and Leda* with one of his crutches 'just as if he were knocking down a bullock'. Adams merely repeated, 'I have nothing to say' when asked why he did it. The magistrate deliberated for a while before realising that he was empowered only to hear cases of criminal damage when the damage in question amounted to under £5, so Adams was duly remanded to a higher court. Hardwick did not have to deliberate before deciding what to do with James Smyth, however; the man was sent straight to Newgate prison next door to the Old Bailey to await trial.

At that trial four days later Smyth's counsel Mr Wilde did his best to argue that the real reason Smyth went back into the parlour was to fetch his hat and coat rather than to get rid of the evidence, but the suggestion had the smack of desperation about it. Sergeant Mount told the court how together the three men had set up the 'sting' operation, and the prosecution then produced a fourth witness. The prisoner's brother-in-law went into the witness box and was shown one of the threatening letters. He confirmed that it was his relative's handwriting. 'Have you been on intimate terms with him?' he was asked. 'Formerly, but not very lately,' Smyth's relative replied.

The prisoner had mentioned his sister, the witness's wife,

in another letter he sent to Robinson, but this one, written in Newgate, struck a rather different tone to his earlier communications with the butler. Begging Robinson to drop the prosecution and admitting his accusations were 'completely trumped up', he said his conviction would cause the death of his only sister and prevent his marriage to a respectable young lady of some property. Robinson, however, was in no mood for forgiveness and nor was the judge when, following the inevitable guilty verdict, it came to sentencing. Referring to the 'diabolical' way that Smyth had chosen to stigmatise his innocent victim of such a repugnant crime, Mr Justice Coleridge ordered the prisoner to be transported for twenty years.

In its report on the case, the *Kentish Independent* referred to a second indictment that the authorities had decided to drop, but which would have further shown the 'systematic course of villainy that Smyth had carried on for some time'. Even so, the paper could not help but comment that the convicted man had somewhat pleasing manners.

James Smyth turned out to be an assumed name. While he was being held in Newgate, the governor Mr Cope discovered that the man being held in custody had in fact been baptised John Bodle, the same John Bodle 'in whose welfare the public took some interest a few years since', said the *News of the World*. His last known appearance in court had been in December 1833 when he was acquitted of murdering his grandfather.

After the Maidstone trial, idleness and a tendency to squander his money had soon caused Young John's shop in Bishopsgate Street to fail, as the Shoreditch coffee shop had failed in 1831. Mary and Thomas Andrews had then taken him in and given him a job in their Clerkenwell coffee shop, but he had repaid them by breaking into Mary's strongbox and disappearing with her life savings (£80), clearly at that time not overly concerned about the health and happiness of his only sister. The couple had heard nothing more of him until the Robinson trial.

Young John's activities in the years between running away from Clerkenwell and his appearance at the Old Bailey are a mystery, although a possible clue to at least one event can be found in the archives of the Eastern Division of Sussex County Assizes. On 24 February 1836, a John Bodle aged twenty-five and described as well educated, stood trial for larceny. That time he was acquitted.

On 20 February 1844, two weeks after the Old Bailey trial, *The Times* carried a small news story under the headline 'The Murder at Plumstead'. 'We understand that John Bodle who was found guilty on Tuesday last and sentenced to 20 years' transportation for extorting money from Lord Abingdon's butler under threat of making a diabolical charge, and who about 10 years ago was tried at Maidstone for the murder of his grandfather by poison, on which occasion he was acquitted, has confessed since his confinement in Newgate that he did commit that murder and that no one but himself was concerned in the horrid transaction.'

Under the double jeopardy rule Young John, once acquitted, could not be tried again for the same offence, and besides, if he retracted the confession there would be no more proof of his guilt then than there had been in 1833. His mother Catherine had died four years earlier at the age of fifty-two in her small rented cottage in Plumstead village. Catherine, like her husband Middle John, was said to have made a deathbed statement about the murder. Perhaps she confided in her nephew, the simple-minded Henry Perks, who was with her when she died. The rumours circulating in Plumstead and Woolwich about Mrs Bodle's last words were vague but according to one resident: 'It was generally understood that she exonerated her husband from the guilt and threw it upon her errant and unnatural son, and she probably explained her own part in the matter, which was not quite free from suspicion.'

There is one final mystery to the Bodle story. Under the terms of old George's will, the proceeds of the sale of land and property prompted by the death of Middle John were to be split equally between his three children. The soft-hearted, industrious Mary Andrews might have received her share, but her brother George lost out somehow. Perhaps in addition to the money he had received, George had also inherited Middle John's profligacy and allowed his entire fortune to slip through his fingers almost as soon as he touched it. Whatever the reason, George was to spend the rest of his working life as an agricultural labourer, drifting from one small Plumstead cottage to another before dying unmarried at the age of seventy-nine in the Woolwich workhouse, the fate that Mary Marsh had so narrowly escaped.

On 1 March Young John Bodle was transferred from Newgate to Millbank Penitentiary on the banks of the Thames, where he remained for several weeks.

By the time Young John arrived, Millbank had been reduced to a holding jail where prisoners were assessed and then moved on, the juveniles to the new Parkhurst prison on the Isle of Wight, those adults considered to possess more moral fibre than the usual run of offenders to Pentonville, and everyone else to a prison hulk or ship for transportation. When the committee responsible for running Millbank as a reformatory handed over power to the new regime, they wrote to the Home Secretary warning that 'a vigorous system will be found necessary for the maintenance of order among criminals of so depraved and desperate a character as the male transports are evidently expected to be'. The authorities wasted little time deciding whether Young John belonged among the depraved and desperate. On this occasion his pleasing manners impressed no one. Six weeks after arriving at Millbank he was led in chains through the front gates of the jail, down the riverbank steps, then put on a boat and taken eastwards along the Thames, back to his old stomping ground at Woolwich, where he was transferred to a ship bound for Australia.

On 22 April 1844, the barque *Maria Somes*, hired by the British government from its commercial owners, set sail for Van Diemen's Land with 264 male prisoners on board. Three of the convicts had faced a court martial but the rest had been tried at assizes and quarter sessions across England, Scotland and Wales, as well as at the Old Bailey. Most had committed the sort of petty offences that occupied Sergeant Mount for most of his working day: Charles Moles, aged eighteen, had stolen a donkey worth £2 from a brewery and sold it to a farmer for a sovereign, while twenty-three-year-old Edward Pooley had made off with a pair of boots from a shop. Both had previous form and both were sentenced to transportation for seven years. Francis Murray's crime was the more serious: 'feloniously breaking and entering the dwelling-house of Peter Crocker, and stealing therein two china ornaments, value one shilling; two glass jugs, 6s; two pairs of ear-rings, 6s; one breast-pin, 3s 6d; two pictures and frames, 5s; and one pair of snuffers, 4d'. He was off for ten years. A few men had received fourteen years but Young John's twenty-year stretch was easily the longest.

On 30 July, after ninety-six days at sea during which two of the convicts died, the men disembarked at their destination. 'Among the prisoners is that well-known character in Plumstead and Woolwich named John Bodle,' reported the *John Bull* magazine. While the *Maria Somes* was moored off the Woolwich dockyard awaiting her full load, some of Young John's relatives applied for permission to visit him before he left the country. There is no record of who they were or whether that permission was ever granted, but this was the last that was heard of the man whose trial had so captivated the nation. Once he reached Australia, John Bodle, along with his alter ego James Smyth, as the official records still listed him, seemed to vanish without a trace.

Notes

The Fell Spirit of the Borgias

Bulwer-Lytton, Edward, *Lucretia; or, the Children of the Night* (2nd edn), London: G. Routledge & Sons, 1889.
The Times, 17 December 1846, p. 7.

Chapter 1: The Big Square House in the Village

Main narrative and general description of Plumstead and Bodles' farm: Vincent, W. T., *The Records of the Woolwich District*, Woolwich: J. R. Jackson, 1888–90; and contemporary national and local newspaper reports of November and December 1833, but especially *The Times*, the *Morning Post*, the *Morning Chronicle*, the *Standard*, the *York Herald*, the *Royal Cornwall Gazette*, the *Leicestershire Chronicle*, the *Maidstone Journal*, the *Maidstone Gazette*.

Births, baptisms, marriages and deaths are from the parish records of St Nicholas, Plumstead; St Mary Magdalene, Woolwich; and St Margaret, Lee, Lewisham.

Property and land ownership: Land Tax Register, National Archives IR 29/17/304, www.kentarchaeology.org.uk/Research/Maps/NOC/02.htm, accessed 19 March 2013.

Plumstead parish churchwardens' accounts, 1820s, Greenwich Heritage Centre.

Philp, Robert Kemp, *The Dictionary of Daily Wants*, London: Houlston and Wright, 1859.

Plumstead parish poor rates and accounts, 1833, Greenwich Heritage Centre.

Plumstead parish poor rates and accounts, 1834–36, Greenwich Heritage Centre.

Description of Young John: National Archives PCOM 2/208; HO 77/51.

National Archives, Wills and Letters of Administration, PROB 11/1826/46.

Price of coffee and sugar: Report from the Select Committee on Agriculture: with the minutes of evidence, appendix and index, 1833. Evidence given in June 1833 by George Porter, Board of Trade statistician.

Chapter 2: A Great Chain and Dependency of Things

Mead, R., *A Mechanical Account of Poisons*, London: Brindley, 1702.

www.toxicology.org/gp/aboutsot.asp, last accessed 19 March 2013.

Witthaus, R. A., *Manual of Toxicology*: reprinted from Witthaus' and Becker's *Medical Jurisprudence, Forensic Medicine and Toxicology*, New York: William Wood, 1911.

Wise, T. A., *Commentary on the Hindu System of Medicine*, London: Smith, Elder and Co., 1845.

Levy, J., *Poison: A Social History*, Stroud: The History Press, 2011.

Macinnis, P., *Poisons: From Hemlock to Botox and the Killer Bean of Calabar*, Skyhorse Publishing Inc., 2011.

Whorton, J. C., *The Arsenic Century*, Oxford: Oxford University Press, 2010.

Emsley, J., *The Elements of Murder*, Oxford: Oxford University Press, 2006.

Weeks, A. (ed. and trans.), *Paracelsus (Theophrastus Bombastus von Hohenheim, 1493–1541): Essential theoretical writings*, Leiden and Boston: Brill, 2008.

Borzellecal, J. F., 'Paracelsus: herald of modern toxicology', *Toxicological Sciences*, 53(1), 2000.

Brodie, B., 'Experiments and observations on the different modes in which death is produced by certain vegetable poisons', *Phil. Trans. R. Soc.*, 101: 337–46, London, 1811.

Brodie, B., 'Further experiments and observations on the action of poisons on the animal system', *Phil. Trans. R. Soc.*, 102: 205–27, London, 1812.

Chapter 3: Death by Toad or Insect

Main narrative is based on contemporary national and local newspaper reports, November and December 1833, but especially *The Times*, the *Morning Post*, the *Morning Chronicle*, the *Standard*, the *York Herald*, the *Royal Cornwall Gazette*, the *Leicestershire Chronicle*, the *Maidstone Journal*, the *Maidstone Gazette*.

Local detail: *Maidstone Journal, Maidstone Gazette*, November 1833.

George Bodle's house: Vincent, W. T., *The Records of the Woolwich District*, Woolwich: J. R. Jackson, 1888–90.

Layout of village: Map of Plumstead, 1869, Greenwich Heritage Centre.

Dates of births, baptisms, marriages and burials: parish records of St Nicholas, Plumstead.

Hempel, S., *The Medical Detective: John Snow, cholera and the mystery of the Broad Street pump*, London: Granta, 2007.

Whorton, J. C., *The Arsenic Century*, Oxford: Oxford University Press, 2010.

Letheby, H., 'On the probability of confounding cases of arsenical poisoning with those of cholera', *Pharmaceutical Journal*, 8: 237, 1848–49.

Nriagu, J. O., 'Arsenic poisoning through the ages', in Henke, K. R. (ed.), *Arsenic: Environmental Chemistry, Health Threats and Waste Treatments*, Chichester: John Wiley, 2009.

Witthaus, R. A., *A Manual of Toxicology*: reprinted from Witthaus' and Becker's *Medical Jurisprudence, Forensic Medicine and Toxicology*, New York: William Wood, 1911.

Levy, J., *Poison: A Social History*, Stroud: The History Press, 2011.

Arlidge, J. T., *The Hygiene Diseases and Mortality of Occupations*, London: Percival, 1892.

Select Committee of the House of Lords on the Sale of Poisons etc. Bill, (HL) PP 1857 (2) XII Minutes of Evidence 658.

The Times, 6 January 1858, p. 6.

The Times, 9 January 1858, p. 11.

Keynes, M., 'Did Napoleon die from arsenical poisoning?', *The Lancet*, 344(8917): 276, 1994.

Emsley, J., *The Elements of Murder*, Oxford: Oxford University Press, 2006. Smith, C. S. and Hawthorn, J. G. (trans.) *Mappae Clavicula, A little key to the world of medieval techniques*, Philadephia: American Philosophical Society, 1974.

Levey, M., *Early Arabic Pharmacology: An introduction based on ancient and medieval sources*, Leiden: Brill, 1973.

Mervyn Madge, A. G., 'Murders and the detection of arsenic', *Pharmaceutical Historian*, 15: 2, 1985.

Chapter 5: There's Not Much in Dying

Groom, N., 'Chatterton, Thomas (1752–1770)', *Oxford Dictionary of National Biography*, Oxford University Press, 2004.

Whorton, J. C., *The Arsenic Century*, Oxford: Oxford University Press, 2010.

Emsley, J., *The Elements of Murder*, Oxford: Oxford University Press, 2006.

Watson, K., *Poisoned Lives*, London and New York: Hambledon and London, 2004.

Witthaus, R. A., *A Manual of Toxicology*, New York: William Wood, 1911.

Witthaus, R. A. and T. C. Becker, *Medical Jurisprudence, Forensic Medicine and Toxicology*, New York: William Wood, 1894–96.

Horne, R. H., *Household Words*, 4: 277, 1851–52.

Parliamentary Papers XXVII. 2 (1862) Fourth Annual Report of the Medical Officer of Health to the Privy Council for 1861 appendix 5, p. 195; PP XXVIII. (1864) Sixth ARMOHPC for 1863, p. 81.

Parliamentary Papers XXVIII. (1864) Sixth Annual Report of the Medical Officer of Health to the Privy Council for 1863, p. 81; appendix 14, pp. 459–60.

Kerr, D., *Forensic Medicine: A text-book for students and a guide for the practitioner*, London: Adam & Charles Black, 1946.

Christison, R., *Life of Sir Robert Christison 1797–1882*, edited by his sons, Edinburgh and London: W. Blackwood and Sons, 1885–86.

The Times, 11 November 1840.

Orfila, M., *A General System of Toxicology, or, a Treatise on Poisons, Drawn from the Mineral, Vegetable, and Animal Kingdoms, Considered as to their Relations with Physiology, Pathology and Medical Jurisprudence*, translated from the French by J. A. Waller, London: E. Cox, 1816–17.

Orfila, M., *A General System of Toxicology*, London: E. Cox, 1821.

Watts, E., 'Poison, proof and a professor: Sir Robert Christison's work in medical jurisprudence in Edinburgh, 1822–1855', Dissertation, 1994.

Christison, R., *A Treatise on Poisons, in Relation to Medical Jurisprudence, Physiology, and the Practice of Physic*, Edinburgh: A. C. Black, 1836.

Crowther, A., 'The toxicology of Robert Christison', in J. R. Bertomeu-Sánchez and A. Nieto-Galan (eds), *Chemistry, Medicine and Crime: Mateu J. B. Orfila (1787–1853) and his times*, Sagamore Beach, MA: Science History Publications, 2006.

Coley, N., 'Alfred Swaine Taylor, MD, FRS (1806–1880)', *Medical History*, 35: 409–27, 1991.

Medical Times and Gazette, 12 June 1880, p. 642.

British Medical Journal, 12 June 1880, p. 905.

The Times, 20 July 1842.

Witthaus, R. A., *A Manual of Toxicology*, New York: William Wood, 1911.

Swaine Taylor, A., *A Manual of Medical Jurisprudence*, London: J. Churchill, 1846.

The Offences Against the Person Act 1861, sections 22, 23 and 24.

Chapter 6: A Great Degree of Inquietude

Bodle narrative is based on contemporary national and local newspaper reports, November and December 1833, but especially *The*

Times, the *Morning Post*, the *Morning Chronicle*, the *Standard*, the *Maidstone Journal*, the *Maidstone Gazette*.

Orfila, M., *Directions for the Treatment of Persons who have taken Poison, and those in a State of Apparent Death*, London: Longman, Hurst, Rees, Orme, and Brown, 1820.

Smith, J. G., *The Principles of Forensic Medicine*, London: Thomas and George Underwood, 1824.

Smith, J. G., *An Analysis of Medical Evidence: Comprising directions for practitioners, in the view of becoming witnesses in Courts of Justice; and an appendix of professional testimony*, London: T. and G. Underwood, 1825.

Smith, T. and H. Smith, 'On an antidote at once for prussic acid, antimony and arsenic', *American Journal of Pharmacy*, 38: 16, 1866.

Beck, T. H., *Elements of Medical Jurisprudence*, London: John Anderson et al., 1825.

The Lancet, II: 1833.

Parr, B., *The London Medical Dictionary*, vol. 1, London: J. Johnson et al., 1809.

Male, G. E., *Elements of Medical Jurisprudence*, London: T. and G. Underwood, 1816.

Report from the Select Committee on the Education and Practice of the Medical Profession in the United Kingdom, Royal College of Physicians, Royal College of Surgeons and Society of Apothecaries, with minutes of evidence, appendices and indices, 1834.

The Lancet, I: 327, 1834–35.

The Lancet, I: 4, 1825–26.

www.apothecaries.org/society/our-history

Jones, R., 'Apothecaries, physicians and surgeons', *Br. J. Gen. Pract.*, 1(56): 24, 232–3, March 2006.

Cope, Z., *The Royal College of Surgeons of England: A history*, London: Anthony Blond, 1959.

www.rcplondon.ac.uk/about/history, last accessed 19 March 2013.

Society of Apothecaries Qualifications Entry Books 1823–26; Society of Apothecaries Register of Apprentice Bindings 1694–1836, MS8207.

Royal College of Surgeons List of Members, 1830.

Loudon, I. S. A., *Medical Care and the General Practitioner, 1750–1858*, Oxford: Clarendon Press, 1986.

Loudon, I. S. A., 'The origin of the GP', *Journal of the Royal College of GPs*, 33: 13–18, 1983.

Munks Roll (of the members of the Royal College of Physicians), vol. II, p. 399.

Carlyle, E. I., 'Sutton, Thomas (1767?–1835)' (rev. Anita McConnell), *Oxford Dictionary of National Biography*, Oxford University Press, 2004.

Sutton, T., *Tracts on Delerium Tremens, on Peritonitis, and on some Other Internal Inflammatory Affections, and on Gout*, London: Underwood, 1813.

Sutton, T., 'An account of some cases of puerperal fever with their treatment', *Edinburgh Medical and Surgical Journal*, 9: 318, 1813.

Minute Book of the Governors of the Kent Dispensary and the Monthly Committee, London Metropolitan Archives H05/M/A/01/001 5 June 1828–14 Aug. 1837. Minute Book of the Governors of the Kent Dispensary. London Metropolitan Archives H05/M/A/01/002 21 Sept. 1837–25 Jan. 1844.

Paris, J. Ayrton and J. S. M. Fonblanque, *Medical Jurisprudence*, London: W. Phillips et al., 1823.

Old Bailey Sessions Papers, 14 January 1789.

Maidstone Journal, 12 November 1833.

Chapter 7: Corroborative Proof as to the Deleterious Article

Bodle narrative is based on contemporary national and local newspaper reports, November and December 1833, especially *The Times*, the *Morning Post*, the *Morning Chronicle*, the *Standard*, the *Maidstone Journal*, the *Maidstone Gazette*.

Christison, R., *A Treatise on Poisons, in Relation to Medical Jurisprudence, Physiology, and the Practice of Physic*, Edinburgh: A. & C. Black, 1836.

Addison, T. and J. Morgan, *An Essay on the Operation of Poisonous Agents upon the Living Body*, London: Longman & Co., 1829.

Hempel, S., *The Medical Detective: John Snow, cholera and the mystery of the Broad Street pump*, London: Granta, 2006.

The Lancet, I: 606, 1833.

The Lancet, II: 281, 1832–33.

Watson, K., *Poisoned Lives*, London and New York: Hambledon and London, 2004.

Forbes, T. R., *Surgeons at the Bailey: English forensic medicine to 1878*, Yale: Yale University Press, 1985.

Beck, T. H., *Elements of Medical Jurisprudence*, London: John Anderson et al., 1825.

Crowther, A. and B. White, *On Soul and Conscience: The medical expert and crime*, Aberdeen: Aberdeen University Press, 1988.

Bellot, H. H., *University College London 1826–1926*, London: University of London Press, 1929.

Ward, J., 'Smith, John Gordon (1792–1833)', *Oxford Dictionary of National Biography*, Oxford University Press, 2004.

The Times, 9 March 1829.

Christison, R., *A Treatise on Poisons, in Relation to Medical Jurisprudence Physiology and the Practice of Physic*, Edinburgh: A. & C. Black, 1836.

Christison, R., *Edinburgh Medical Journal*, xxxi: 236–50, April 1829.

The Times, 17 September 1833.

Swaine Taylor, A., *A Manual of Medical Jurisprudence*, London: J. Churchill, 1846.

Smith, J. G., *The Principles of Forensic Medicine*, London: Thomas and George Underwood, 1824.

Old Bailey Sessions Papers, February 1835.

The Lancet, 1: 47, 1833–34.

Chapter 8: Those Low Incompetent Persons

Bodle narrative is based on contemporary national and local newspaper reports of the inquest and trial, November and December 1833, but especially *The Times*, the *Morning Post*, the *Morning Chronicle*, the *Standard*, the *Maidstone Journal*, the *Maidstone Gazette*.

Ludlow, B., 'Plumstead: A 19th [sic] suburb of Woolwich's industrial and military might', www.ideal-homes.org.uk/case-studies/plumstead, last accessed 19 March 2013.

Plumstead Population of Kent Parish Census 1801 to 1921. Page, W. (ed.), *The Victoria History of the County of Kent*, vol. 3, London: St Catherine Press, 1932.

Property and land ownership: Land Tax Register National Archives: IR 29/17/304.

Pigot's National & Commercial Directory 1832/33/34, Pubs and Brewers of Kent; http://freepages.genealogy.rootsweb.ancestry. com/~mrawson/pubsdir1.html#maidstone, last accessed 19 March 2013.

Plumstead parish poor rates and accounts, 1830/1833/1834–36, Greenwich Heritage Centre.

1841 census return. National Archives HO107/484/19.

Coroners' Jury Members: National Archives ASSI 94/2164.

Charles Carttar succeeds Joseph: National Archives C 202/221/29.

Knapman, P., 'The Crowner's quest', *Journal of the Royal Society of Medicine*, 86, December 1993.

Grindon, Joseph B., *The Law and Duties of the Important Office of Coroner*, Bristol: Baldwin, 1822.

The Times, 9 March 1825, p. 3.

Ayrton Paris, J. and J.S.M. Fonblanque, *Medical Jurisprudence*, London: W. Phillips et al., 1823.

Dickens, C., *Bleak House*, London: Wordsworth Editions, 1993.

Wills, W. H., 'A coroner's inquest', *Household Words*, 27 April 1850, p. 109.

The Lancet, II, 1827–28.

Smith, J. G., *An Analysis of Medical Evidence*, London: T. and G. Underwood, 1825.

The Times, 2 May 1849, p. 8.

Chapter 9: A Very Active Constable

Main narrative is based on contemporary national and local newspaper reports, November and December 1833, especially *The Times*, the *Morning Post*, the *Morning Chronicle*, the *Standard*, the *Maidstone Journal*, the *Maidstone Gazette*.

http://www.met.police.uk/history/metropolitan historical collection. html, last accessed 20 March 2013.

Greenwich, Woolwich and Deptford Gazette, 11 January 1834.

West Kent Guardian, 4 January 1840.

The Times, 1 January 1823.

Christison, R., *A Treatise on Poisons*, Edinburgh: A. & C. Black, 1836.

Swaine Taylor, A., *A Manual of Medical Jurisprudence*, London: J. Churchill, 1846.

Anonymous, *The Complete Vermin-killer*, London: Fielding and Walker, 1777.

Bartrip, P., 'A pennurth of arsenic for rat poison', *Medical History*, 36: 53–69, 1992.

Pharmaceutical Journal, i: 329, 1841–42.

Bell, J. and T. Redwood, *Historical Sketch of the History of Pharmacy*, London: Pharmaceutical Society of Great Britain, 1880.

The Lancet, I: 18, 1845.

Old Bailey Sessions Papers, 6 January 1845.

Parliamentary Papers 1819. A Bill for Establishing Regulations for the Sale of Poisonous Drugs and for the Better Preventing the Mischiefs arising from the inattention or neglect of Persons vending the same.

Thompson, B. J. H., 'Notes on Dr Thomas Goulard's treatise on the effects and various preparations of lead ...', *Proceedings of the Royal Society of Medicine*, xxxi: 1435, 1938.

The Lancet, I: 589, 642, 693, 1846.

The Times, 20 May 1846, p. 6.

The Times, 22 June 1846, p. 4.

British Medical Journal, 304, 1857.

Sprigge, S. S., *The Life and Times of Thomas Wakley*, London, New York and Bombay: Longmans, Green, 1897.

The Times, 23 and 26 August 1830.

Old Bailey Sessions Papers, 30 October 1830 and 17 February 1831.

Long, J. St John, *A Defence of John St John Long Esq in the Cases of the Late Catherine Cashin and Mrs Colin Campbell Lloyd Founded upon the Evidence Against Him*, London: Chapple, 1831.

Richardson, R., 'Coroner Wakley: two remarkable eyewitness accounts', *The Lancet*, 358(9299): 2150–54, 2001.

Minute Book of the Governors of the Kent Dispensary, London Metropolitan Archives, H05/M/A/01/001.

The Lancet, II: 401, 1828–29.

Chapter 10: The Introduction of Irritating Matter

Main narrative is based on contemporary newspaper reports, November and December 1833, in *The Times* and the *Maidstone Journal*.

Watson, K. D., 'Medical and chemical expertise in English trials for criminal poisoning, 1750–1914', *Med. Hist.*, 50(3): 373–90, 1 July 2006.

Forbes, T. R., *Surgeons at the Bailey: English forensic medicine to 1878*, Yale: Yale University Press, 1985.

Old Bailey Sessions Papers, 30 August 1786.

Crawford, C., 'Medicine and the law', in W. Bynum and R. Porter (eds), *Companion Encyclopedia of the History of Medicine*, London and New York: Routledge, 1993.

Smith, J. G., *The Claims of Forensic Medicine, being the introductory lecture delivered in the University of London on May 11, 1829*, London: John Taylor, 1829.

Smith, J. G., *An Analysis of Medical Evidence: Comprising directions for practitioners, in the view of becoming witnesses in Courts of Justice; and an appendix of professional testimony*, London: T. and G. Underwood, 1825.

Hunter, W., 'On the uncertainty of the signs of murder in bastard children', in J. Dowson, *An Introduction to the Study and Practice of Medicine*, London: Longman, Rees, Orme, 1834.

Old Bailey Sessions Papers, 16 April 1795.

Act to Provide for the Attendance and Remuneration of Medical Witnesses at Coroners' Inquests: 11 6 & 7 Wm IV c 89.

Vincent, W. T., *The Records of the Woolwich District*, Woolwich: J. R. Jackson, 1888–90.

Pigot's Directory, 1827, 1838, 1840.

Jacyna, L. S., 'Solly, Samuel (1805–1871)', *Oxford Dictionary of National Biography*, Oxford University Press, 2004.

Pearce, J. M. S., 'A note on scrivener's palsy', *J. Neurol. Neurosurg. Psychiatry*, 76: 513, 2005.

Swaine Taylor, A., *A Manual of Medical Jurisprudence*, London: J. Churchill, 1846.

Morning Post, 8 November 1833.

Chapter 11: I Never Saw Two Things in Nature More Alike

Main narrative is based on contemporary newspaper reports, November and December 1833, particularly *The Times* and the *Maidstone Journal*.

James, Frank A. J. L., 'Farqaday, Michael (1791–1867)', *Oxford Dictionary of National Biography*, Oxford University Press, 2004.

James, F. (ed.), *The Correspondence of Michael Faraday*, vol. 1, London: Institution of Electrical Engineers, 1991.

James, Frank A. J. L., 'Marsh, James (1794–1846)', *Oxford Dictionary of National Biography*, Oxford University Press, 2004.

Ordnance Minutes, National Archives: WO47/1,841, p. 12,747.

Hogg, O. F. G., *The Royal Arsenal, Woolwich*, vols 1 & 2, Oxford: Oxford University Press, 1963; *Morning Post*, 8 June 1837.

1841 census return.

Parish baptism records, St Mary Magdalene, Woolwich.

Pigot's Directory, Woolwich, 1840.

Greenwich Heritage Centre: miscellaneous papers: 'ephemera'.

Hasted, E., *The History and Topographical Survey of the County of Kent: Volume 1*, 1797, pp. 441–54.

Vincent, W. T., *The Records of the Woolwich District*, Woolwich: J. R. Jackson, 1888–90.

Vestry Minutes for the Parish of Plumstead, 1839, Greenwich Heritage Centre.

Ordnance Minutes, National Archives: WO 47/978.

Ordnance Minutes, National Archives: WO 44/295, letter 17 August 1846.

Army records, National Archives: WO 97/1174/1; WO 116/125.

Pigot's Directory, Woolwich, 1840.

1851 census returns piece, 1589, folio 212, p. 24.

RSA Committee minutes 1822–23: PR/GE/112/12/64.

Clerke, A. M., 'Barlow, Peter (1776–1862)' (rev. Iwan Rhys Morus), Oxford Dictionary of National Biography, Oxford University Press, 2004.

RSA Transactions 1823, vol. 41.

Ordnance Minutes, National Archives: WO 44/295, report 26 August 1846.

Anon., The Domestic Chemist, London: Bumphus and Griffin, 1831.

The Times, 13 August 1821.

Christison, R., A Treatise on Poisons, Edinburgh: Adam Black, 1832.

Anon., The Tryal of Mary Blandy Spinster for The Murder of her Father, Francis Blandy, Gent ..., London: John and James Rivington, 1753.

Roughead, W. (ed.), Trial of Mary Blandy, Edinburgh, 1914.

'The trial of Miss Blandy for poisoning her father', The Gentleman's and London Magazine, and Monthly Chronologer, March 1752, pp. 136–46.

Wax, P., 'The origins of analytical toxicology, arsenic detection, and the trials of Mary Blandy and Marie Lafarge', Mithridata, 12(2): 24, 2002.

Marshall, J., Five Cases of Recovery from the Effects of Arsenic ..., London: C. Chapple, 1815.

Old Bailey Sessions Papers, April 1815, ref: t18150405-18.

Smith, J. G., Hints for the Examination of Medical Witnesses, London: Longman, Rees, Orme, Brown, and Green, 1829.

Smith, J. G., The Claims of Forensic Medicine, being the introductory lecture delivered in the University of London on May 11, 1829, London: John Taylor, 1829.

Watkins, J., The Important Results of an Elaborate Investigation into the Mysterious Case of Elizabeth Fenning ..., London: Hone, 1815.

The Times, 31 March 1815.

Watson, K., *Poisoned Lives*, London and New York: Hambledon and London, 2004.

Chapter 12: She Would Not Risk Her Soul into Danger

The Times, the *Morning Chronicle*, the *Maidstone Gazette*, the *Maidstone Journal*, all of 12 November 1833

Kentish Independent, December 1843.

Greenwich Gazette, 3 April 1834.

National Archives: HO 27/25 p247; National Archives: ASSI 94/1873.

Chapter 13: Oh My Poor Mother

Main narrative *The Times*, the *Morning Chronicle*, the *Morning Post*, 15 November 1833.

Eastoe, J. with R. Goodman, *Victorian Pharmacy Remedies and Recipes*, London: Pavilion, 2010.

The Lancet, II: 828, 1838–39.

The Lancet, I: 744, 1829–30.

The Lancet, II: 175, 1833–34.

The Lancet, II: 402, 1837–38.

Rennie, J., *A New Supplement to the Latest Pharmacopoeias of London, Edinburgh, Dublin and Paris*, London: Baldwin and Cradock, 1837.

Christison, R. and R. E. Griffith, *A Dispensatory; or, Commentary on the Pharmacopoeias of Great Britain (and the United States)*, Philadelphia: Lea and Blanchard, 1848.

Parr, B., *The London Medical Dictionary*, vol. 1, London: J. Johnson et al., 1809.

Begbie, J., *Arsenic, its physiological and therapeutical effects*, Edinburgh: Murray and Gibb, 1858.

Parish baptism records: St Mary Magdalene, Woolwich.

Morning Chronicle, 23 November 1833.

Chapter 14: From the Very Brink of Eternity

Main narrative national and regional newspaper accounts, mainly *The Times*, the *Morning Chronicle*, the *Morning Post* and the *Standard*, 13 and 14 December 1833.

Trial papers, Kent Winter Assizes: National Archives ASSI 94/2164.

Boase, G. C., 'Gaselee, Sir Stephen (1762–1839)' (rev. Sinéad Agnew), *Oxford Dictionary of National Biography*, Oxford University Press, 2004.

Corsi, P., 'Baden-Powell (1796–1860)', *Oxford Dictionary of National Biography*, Oxford University Press, 2004.

Kent County Archives PCM/1.

Drew, J. M. L., *Dickens the Journalist*, Palgrave Macmillan, 2003.

House, H., *The Dickens World*, London: Oxford University Press, 1960.

The Times, 20 April 1837.

Morning Chronicle, 19 December 1823.

Kent County Archives Q/GAC/2.

Hastings, P. and I. Coulson, Kent County Council website, http://www.hereshistorykent.org.uk/choosearticle.cfm, last accessed 20 March 2013.

Howard, J., *The State of the Prisons in England and Wales*, London: Warrington, 1777.

Chalklin, C. W., *English Counties and Public Building 1650–1800*.

Melling, E. (ed.), *Kentish Sources, 6: Crime and Punishment*, Maidstone: Kent County Council, 1969.

Kent County Archives, PCM/2 and PCM/3.

Trial papers, Kent Winter Assizes: National Archives ASSI 94/2164.

Chelmsford Chronicle, 27 December 1833.

John Bull, 30 December 1833, p411.

Essex Standard and *Colchester and County Advertiser*, 21 December 1833.

Chapter 15: The Sequel of These Proceedings

Main narrative national and regional newspaper reports: *The Times*, the *Morning Chronicle*, the *Morning Post*, the *Standard*.

John Bull, 30 December 1833, p411.

The Spectator, January 1834.

Kentish Gazette, 24 December 1833.

Plumstead parish poor rates and accounts, 1833, Greenwich Heritage Centre.

Chelmsford Chronicle, 27 December 1833.

Kentish Gazette, 21 January 1834.

Plumstead parish poor rates and accounts, 1834–36, Greenwich Heritage Centre.

Vincent, W. T., *The Records of the Woolwich District,* Woolwich: J. R. Jackson, 1888–90.

Parliamentary Papers 1835. Reports from the Select Committee of the House of Lords appointed to enquire into the changes of the county rates in England and Wales ..., pp. 297–9.

Marsh, J., 'Account of a method of separating small quantities of arsenic from substances with which it may be mixed. By James Marsh, Esq. of the Royal Arsenal, Woolwich. (Communicated to the Society of Arts of London)', *Edinburgh New Philosophical Journal,* April–Oct. 1836.

Whorton, J. C., *The Arsenic Century,* Oxford: Oxford University Press, 2010.

Bertomeu-Sánchez, J. R., 'Sense and sensitivity', in J. R. Bertomeu-Sánchez and A. Nieto-Galan (eds), *Chemistry, Medicine, and Crime: Mateu J. B. Orfila (1787–1853) and his times,* Sagamore Beach, MA: Science History Publications, 2006.

RSA Committee minutes 1822–23, PR/GE/112/12/64; RSA Transactions 1823, vol. 41.

RSA Transactions 1836–37 and 1837–38, vol. 51; Minutes of Committee 1835–36: PR/GE/112/12/77.

Morning Post, 8 June 1836.

Burney, I., *Poison, Detection and the Victorian Imagination,* Oxford: Oxford University Press, 2006.

Crowther, A., 'The toxicology of Robert Christison', in J. R. Bertomeu-Sánchez and A. Nieto-Galan (eds), *Chemistry, Medicine, and Crime: Mateu J. B. Orfila (1787–1853) and his times,* Sagamore Beach, MA: Science History Publications, 2006.

Bertomeu-Sánchez, J. R., 'Sense and sensitivity', in J. R. Bertomeu-Sánchez and A. Nieto-Galan (eds), *Chemistry, Medicine, and Crime: Mateu J. B. Orfila (1787–1853) and his times*, Sagamore Beach, MA: Science History Publications, 2006.

Lafarge case: *The Times, Morning Chronicle*, 15 February–11 August; *The Times*, 8, 10, 14, 15 September; 1 October 1840.

Wax, P., 'The origins of analytical toxicology, arsenic detection, and the trials of Mary Blandy and Marie Lafarge', *Mithridata*, 12(2): 24, 2002.

Lynch, M. H., 'Analysis of Madame Lafarge's trial', *Prov. Med. Surg. J.*, 1(2), 10 October 1840.

New York Times, 8 November 1874.

Reinsch, H., 'On the action of metallic copper on solutions of certain metals, particularly with reference to the detection of arsenic', *Philosophical Magazine*, 19: 480–83, 1848.

RSA Transactions 1836–37 and 1837–38, vol. 51.

Minutes of RSA Committee 1835–36: PR/GE/112/12/77.

Pharmaceutical Journal and Transactions, 1: 278, 1841–42.

Chapter 16: What if the Chymist Should Be Mistaken?

Bulwer-Lytton, Edward, *Lucretia; or, the Children of the Night*, London: Saunders and Otley, 1846.

Daily News, 12 December 1846.

Morning Chronicle, 1 January 1847.

Morning Post, 3 February 1847.

London Medical Gazette, 4: 242–8, 1847.

De Quincey, T., *On Murder Considered as One of the Fine Arts*, Oxford: Oxford University Press, 2006 edn.

Forbes, T. R., *Surgeons at the Bailey*, New Haven, CT: Yale University Press, 1985.

Morley, H., *Household Words*, 13: 221, 1856.

The Leader, 15 December 1855.

The Times, 22 August 1859.

Burney, I., *Poison, Detection and the Victorian Imagination*, Manchester: Manchester University Press, 2006, p.20.

Bynum, W. F., S. Lock and R. Porter (eds), *Medical Journals and Medical Knowledge*, New York: Routledge, 1992.

Lloyd's Weekly, 8 June 1856.

Chadwick, E., *Report on the sanitary conditions of the labouring population of Great Britain. A supplementary report on the results of a special inquiry into the practice of interment in towns*, London: HMSO, 1843.

Lloyd's Weekly, 12 August 1849.

Daily News, 20 August 1849.

The Era, 19 August 1849.

The Times, 19 and 21 September 1846; 5 September 1850; 8 March 1851.

Daily News, 5 September 1846.

The Era, 10 September 1848.

Morning Post, 5 September 1848.

Ipswich Journal, 10 March 1849.

The Times, 22 September 1848.

Watson, K., *Poisoned Lives*, London and New York: Hambledon and London, 2004, p. 45.

Burney, I., *Poison, Detection and the Victorian Imagination*, Manchester: Manchester University Press, 2006, p. 20.

House of Lords Debate, 24 March 1851, Hansard, vol. 115, cols 422–4.

Mill, J. S., *The Letters of John Stuart Mill*, vol. 1, London: Longmans, Green, 1910.

House of Lords Debate, 5 June 1851, Hansard, vol. 117, col. 444.

Bristol Mercury, 18 April 1835.

The Era, 17 February 1850.

Old Bailey Sessions Papers, 4 March 1850.

Daily News, 9 March; 14 March; 18 March 1850.

Medical Times, 23 March 1850.

Best, W. M., *A Treatise on the Principles of Evidence and Practice as to Proofs in Court of Common Laws*, London: Sweet, 1860.

Smethurst case: Old Bailey Sessions Papers, ref. t18590815-785.

Stephen, J. F., *A History of the Criminal Law of England*, vol. 1, Abingdon: Taylor and Francis, 1996.

British Medical Journal, 27 August 1859, p. 702.

The Times, 26 August 1859, p. 9.

Pharm J., 1860–61; 2 (2nd series), p. 337.

Pharm J., 1860–61; 2 (2nd series), p. 475.

British Medical Journal, 3 September 1859, p. 725.

The Times, 22 August 1859.

Chapter 17: The Freezing Influence of Official Neglect

Morning Post, 8 June 1837.

RSA Transactions 1836–37 and 1837–38, vol. 51.

RSA Committee minutes 1836–37 PR/GE/112/12/78.

National Archives WO/47/1,841, p. 12747.

RSA Transactions 1838–39, vol. 52.

RSA Committee papers 1838–39 PR/GE/112/12/80.

Ordnance minutes, National Archives, WO/47/1879, p. 13828; WO/47/1855, p. 3563.

BMD Death certificate: Apr.–Jul. 1846, 05/211.

National Archives PROB/11/2058.

National Archives WO 44/295, Letter 7 August 1846.

National Archives WO 44/295, Petition received 10 August 1846.

National Archives WO 44/295, 28 Aug. 1846, Letter 3 Sept. 1846.

National Archives WO 44/295, Letter 4 September 1846.

National Archives WO 44/295, 5 October 1846.

Morning Chronicle, 16 November 1846.

The Times, 5 November 1846.

The Times, 12 November 1846.

National Archives WO 44/295, Letters 28 October, 2 November 1846.

National Archives WO 44/295, Letters 3, 7, 19 April 1847.

The Examiner, 20 November 1847.

Chapter 18: I Went with a Lie in my Mouth

Greenwich Heritage Centre, Plumstead parish poor rates and accounts 1834–36.

St Nicholas, Plumstead, parish records, burial.

Census returns 1841, piece 461, book 4, folio 7, p. 6.

Old Bailey Sessions Papers, 9 May 1842.

BMD Death certificate: Oct.–Dec. 1843, vol. 5, p. 201.

Vincent, W. T., *The Records of the Woolwich District*, Woolwich: J. R. Jackson, 1888–90.

National Archives PCOM 2/208; HO 77/51.

Old Bailey Sessions Papers, 5 February 1844.

National Archives CRIM 4/241.

National Archives HO 16/8.

National Archives CRIM 5/2.

Old Bailey Sessions Papers, 14 December 1840.

The Times, 16 December 1840.

West Kent Guardian, 10 February 1844.

National Archives PCOM 2/208.

News of the World, 11 February 1844.

National Archives ASSI 31/38.

Census returns 1841, 1851, 1861, 1871, 1881; parish records St Margaret Plumstead, parish records, burial 9 April 1886.

National Archives PCOM 2/21.

Griffiths, A., *Memorials of Millbank*, London: Chapman and Hall, 1884.

www.convictrecords.com.au/convicts/ship-name/maria-somes

Old Bailey Sessions Papers, 1 January 1844.

Old Bailey Sessions Papers, 5 February 1844.

Old Bailey Sessions Papers, 4 March 1844.

John Bull, 16 March 1844.

Kentish Gazette, 19 March 1844.

Main sources for the Bodle story

No official records of the Bodle inquest survive, so the narrative here is taken from national and local newspaper reports, mainly from *The Times, The Morning Post, The Morning Chronicle*, the *Maidstone Journal* and the *Maidstone Gazette*.

Sources for the trials are: National Archives HO 27/25 p247, HO 16/8, PCOM 2/208, PCOM 2/21 and CRIM 4/241; Old Bailey Sessions Papers 5th February 1844, ref t18440205-583.

Other main sources for the Bodle story and for life in Plumstead and Woolwich in 1833 are W. T. Vincent's *The Records of the Woolwich District* and local newspapers, especially the *Maidstone Journal, Maidstone Gazette, Kentish Gazette, Kentish Independent, West Kent Guardian*, and the *Greenwich, Woolwich and Deptford Gazette*.

Acknowledgements

My grateful thanks go to the following people who have so generously given me their time and their expertise, not to mention on occasion their delightful company.

John Slaughter, forensic toxicologist, for his endless patience and skill in reading drafts, answering queries and explaining such a complex subject to such a non-scientist.

Professor Robert P. Chilcott, Head of Toxicology, Department of Pharmacy, University of Hertfordshire, for his knowledge of the history of toxicology, the nineteenth century toxicologists and James Marsh.

Dr John M.T. Ford, MD, MRCGP, DHMSA, for his insights into the history of medical education and the development of the medical profession, and for sharing with me new, as yet unpublished, material from his original research.

Andrew Cunningham, Senior Research Fellow, Department of History and Philosophy of Science, University of Cambridge, for reading the entire final manuscript and providing his usual indispensable advice and encouragement.

Dr Nick Cambridge, medical historian, MD, MRCS, FSA, FRSA, FLS, for dealing with my clinical queries with his customary kindness and good humour.

Joan Craig, Bsc (Pharm), MRPharmS, for her expertise on pharmacology, and for all her interest and encouragement throughout the research and writing process.

David Massa, retired solicitor, for his guidance through the minefield of nineteenth century wills and inheritance law, as well as his enthusiasm and support.

His Honour Graham Boal QC for reading the early draft, and for his advice on the legal definition of poison and the laws of evidence.

Frances Ward and Jonathan Partington of the Greenwich Heritage Centre, for their input on the history of Plumstead and Woolwich, and their help in sourcing material; likewise to Janet Payne, archivist, Worshipful Society of Apothecaries, for a highly informative visit to Apothecaries Hall.

Rebecca Storr, Collections Access Coordinator, Science Museum, for a fascinating visit to see versions of James Marsh's apparatus in the museum storeroom.

Linda and Stanley Slaughter for alerting me to the story of Thomas Chatterton.

Naveed Khokhar for his essential Photoshop skills.

The staff at that wonderful resource The Wellcome Library, as well as those at the National Archives and the British Library, for their helpfulness and expertise.

My agent, the legendary Patrick Walsh, for his usual unfailing acumen, support and friendship.

The team at Weidenfeld & Nicolson, particularly Jessica Gulliver, Sophie Buchan, Hannah Whitaker and Elizabeth Preston, and especially Kirsty Dunseath, for all her enthusiasm and encouragement, and also for what appears to be her ability to mind-read when it comes to getting text to where I want it to be.

And finally the usual grateful thanks to family and friends for all their vital interest and encouragement while I bored on. And on ....

INDEX

blog and newsletter

For exclusive short stories, poems, extracts, essays, articles, interviews, trailers, competitions and much more visit the Weidenfeld & Nicolson blog and sign up for the newsletter at:

www.wnblog.co.uk

Follow us on

 and **twitter**

Or scan the code to access the website*